T0135783

Electromembrane desalination processes for production of low conductivity water

Von der Fakultät Energie-, Verfahrens- und Biotechnik
der Universität Stuttgart zur Erlangung der Würde
eines Doktors der Ingenieurwissenschaften (Dr.-Ing.)
genehmigte Abhandlung

vorgelegt von

Andrej Grabowski

geboren in Nowopokrowskaja, Russland

Hauptberichter: Prof. Dr.-Ing. G. Eigenberger
Mitberichter: Prof. Dr.-Ing. H. Strathmann

Tag der mündlichen Prüfung 27.10.2010

Institut für Chemische Verfahrenstechnik
der Universität Stuttgart

2010

Bibliografische Information der Deutschen Nationalbibliothek

Die Deutsche Nationalbibliothek verzeichnet diese Publikation in der
Deutschen Nationalbibliografie; detaillierte bibliografische Daten sind
im Internet über http://dnb.d-nb.de abrufbar.

D 93

ISBN 978-3-8325-2714-3

Logos Verlag Berlin GmbH
Comeniushof, Gubener Str. 47,
10243 Berlin
Tel.: +49 (0)30 42 85 10 90
Fax: +49 (0)30 42 85 10 92
INTERNET: http://www.logos-verlag.de

Preface

The present work, aimig at the development of electrodialysis with profiled membranes and of new concepts for electrodeionization was funded by the Willy Hager Foundation, Stuttgart. The work was accomplished during my stay from 2001 to 2006 at the Institute for Chemical Processes Engineering (ICVT), Universität Stuttgart. I want to give special thanks to Prof. Heiner Strathmann, who initialized the project and invited me to work in Stuttgart. I'm greatly obliged to him for his steady interest and support of my work and for encouraging discussions. To Prof. Gerhart Eigenberger I'm very grateful for the unique chance to work at ICVT under his management and for his active participation and support during both, "experimental" and "writing" stages of the work.

Many thanks to all colleagues from ICVT, who helped in the work. In particular I want to acknowledge Sven Thate for his investigations of continuous electrodeionization, which created a basis of my work, Gerhard Friedrich for designing much of my equipment, Jochen Kerres and Martin Hein for organic synthesis, Nick Specogna and Holger Aschenbrenner for support with laboratory equipment and several students for the assistance in routine work. Special thanks to Galina Schumski and Inna Kharitonova for membrane characterisation, to Robert Fettig for the help in the laboratory and all three of them for interesting lunch-time communications.

A big acknowledgement to Guiquing Zhang for his active work on the improvement of electrodeionization with bipolar membranes, which he performed at ICVT from 2003 to 2004 under financial support of the "China Scholarship Council".

The production of profiled membranes was possible thanks to the strong support from "Institut für Kunststofftechnik" (Stuttgart) under Prof. Fritz, which provided the equipment for membrane preparations carried out under the highly qualified guidence of Ralf Kaiser.

A big thank to Alexandra Isaeva-Wolf from the "Institut für Mechanische Verfahrenstechnik" (Stuttgart) for measurements with the particle size analyser and gas pyknometer.

It was very encouraging to realise an interest in this work from industrial partners: "FumaTech GmbH" supported us with ion-exchange membranes and polymer solutions, and "Hager + Elsässer GmbH" provided us with ion-exchange resins and Si-analysis.

Using this opportunity, I also like to recognize my colleagues from the Physical Chemistry Department of Kuban State University (Krasnodar, Russia), where I was educated and had a pleasure to work in the atmosphere of a very experienced school in electromembrane science. Some results of ion-exchange resins characterization obtained in Krasnodar have been used in this work for the explanaition of resin's fundamental properties. Particular thanks to my scientific advisers and tutors from Krasnodar Prof. Victor Zabolotskiy, Prof. Nikolai Sheldeshov and Prof. Victor Nikonenko for my scientific and personal development.

Finally, the basic gratitude to my mother for supporting me in my education. And, for sure, the greatest thanks to my beloved wife Ludmila, for her patience and support, without which the work wouldn't been completed.

Montigny le Bretonneux (France), February 2010

Content

VI

List of symbols

Latin symbols

Symbol	Description	Unit
A	area	m^2
a	coefficient	-
a_i	activity of i-component	mol/m^3
C	molar concentration	mol/m^3
D	diffusion coefficient	m^2/s
d_h	hydraulic diameter	m
d_p	diameter of spherical particles	m
E^V	energy consumption per volume of diluate	$W \cdot h/m^3$
F	Faraday constant	96486 $A \cdot s/eq$
G	electric conductance	S
$\widetilde{G}$	electrochemical free Gibbs energy	J
h	thickness	m
i	current density	A/m^2
$\boldsymbol{J}$	molar flux vector	$mol/(m^2 \cdot s)$
K	equilibrium constant	unit depends on reaction order
l	length	m
l_D	entry length of the flow channel	m
L_h	hydrodynamic permeability	$m^2/(s \cdot Pa)$
MW	molar weight	kg/mol
n	coefficient	-
P	pressure	Pa
Q	flow rate	m^3/s
R	universal gas constant	8,31 $J/(mol \cdot K)$
R	electric resistance	Ohm
Re	Reynolds number	-
T	absolute temperature	K
t_i	transport number of an ion i	-
u	electric ion mobility	$m^2 \cdot V^{-1} \cdot s^{-1}$

U	voltage on a module	V
v	linear velocity	m/s
V	volume	m^3
V_m	partial molar volume	m^3/mol
x	linear coordinate	m
z	ion charge number	eq/mol

Greek symbols

Symbol	Description	Unit
$\tilde{\mu}$	electrochemical potential	J/mol
Φ	void fraction (porosity) of a bed	-
Δ	recovery rate	%
δ	effective thickness of Nernst's diffusion layer	m
ε	current efficiency	-
ϕ	electric potential	V
γ	activity coefficient	-
η	dynamic viscosity	Pa·s
φ	volume fraction of a component in a mixture	parts of unit, or vol.%
κ	specific electrical conductivity	S/m or µS/cm
ν	stoichiometric coefficient in the chemical formula	-
π	"pi"	3,14
θ	equivalent fraction of an ion	-
ρ	mass density	kg/m^3
ρ	specific electrical resistivity	MOhm·cm
τ	time	s

Abbreviations

Abbreviation	Description
AAS	Atomic absorption spectrophotometry
AM	Anion-exchange Membrane
^{PC}AM	Anion-exchange Membrane of Protection Compartment
AR	Anion-exchange Resin
BM	Bipolar Membrane
CC	Concentrate Compartment
CEDI	Continuous Electrodeionization
CEDI-BM	Continuous Electrodeionization with Bipolar Membrane
CEDI-BM-Pro	Continuous Electrodeionization with Bipolar Membrane and Protection Compartment
CM	Cation-Exchange Membrane
CR	Cation-Exchange Resin
DC	Diluate Compartment
IC	Ion-Chromatography
IEC	Ion-Exchange Capacity
IER	Ion-Exchange Resin
PC	Protection Compartment
STW	Softened Tap-Water
TDS	Total Dissolved Solids
UPW	Ultra-Pure Water
UV	Ultra-Violet light
PVC	Polyvinylchloride
PE	Polyethylene
PP	Polypropylene
PA	Polyamide
DVB	Divinylbenzene
IEM	Ion-exchange membrane
ED	Electrodialysis
EED	Electro-electrodialysis

Zusammenfassung

Elektromembran-Entsalzungsprozesse für die
Herstellung von Reinwasser mit niedriger Leitfähigkeit

Elektromembran-Entsalzungsprozesse spielen eine wichtige Rolle in der Trinkwasseraufbereitung und Reinstwasserherstellung. Zusammen mit anderen Wasseraufbereitungsprozessen haben sie mit zunehmendem Wassermangel erheblich an Bedeutung gewonnen. Ziel der Arbeit ist es, neue Wege zur Verbesserung und Weiterentwicklung der Elektromembranverfahren zu finden und zu untersuchen.

Am Anfang der Arbeit werden die Eigenschaften sowie die physikalisch-chemischen Grundlagen von Transportprozessen von Ionen in Lösungen und Feststoffelektrolyten beschrieben.

Nachfolgend werden im zweiten Kapitel Elektromembranverfahren mit dem Schwerpunkt Wasseraufbereitung vorgestellt und die Elektrodialyse und kontinuierliche Elektrodeionisation eingehend beschrieben. Als Hauptkomponenten in Elektromembranverfahren werden Eigenschaften und Herstellungsmethoden von Ionenaustauscherharzen und Membranen ausführlich diskutiert.

Elektromembranverfahren können in zwei Gruppen unterteilt werden:

a) Prozesse, die nur auf Ionenmigration basieren und meistens zur Entsalzung von Lösungen mit relativ hohen Konzentrationen eingesetzt werden, wie z.B. die Elektrodialyse.

b) Prozesse, bei denen zusätzlich zur Ionenmigration auch ein Ionenaustausch zur Entfernung von Verunreinigungen eingesetzt wird, wie z.B. die kontinuierliche Elektrodeionisation. Diese Verfahren werden meistens zur Produktion von hochreinem Prozesswasser aus Oberflächen- und Brunnenwasser, das bereits durch andere Methoden vorgereinigt ist, eingesetzt.

Zwei aussichtsreiche Möglichkeiten zur Weiterentwicklung der Elektrodialyse und der kontinuierliche Elektrodeionisation werden im Folgenden als Ziele der Arbeit ausgewählt:

1. Entwicklung eines neuen Elektrodialyseverfahrens mit profilierten Membranen mit dem Ziel der Kostensenkung und der Verbesserung der Produktwasserqualität.

2. Untersuchung der kontinuierlichen Elektrodeionisation mit Bipolarmembranen zur Reinstwasserherstellung mit dem Ziel die Produktwasserqualität weiter zu verbessern.

Dem entsprechend befasst sich das dritte Kapitel mit der Entwicklung von neuen profilierten Membranen und experimentellen Untersuchungen ihrer Leistung unter variablen Betriebsbedingungen. Die Entwicklung von Membranen, die für die Entsalzungsversuche in Frage kommen, basiert auf der Auswahl der Profilgeometrie, des zur Herstellung der Membranen eingesetzten Materials und der eigentlichen Herstellungsmethode.

Für die Auswahl der Profilgeometrie müssen die mechanische Stabilität des Membranstapels, der Druckverlust und die Durchmischung im Kanal zwischen den Membranen berücksichtigt werden. Basierend auf einem Literaturstudium von Forschungsarbeiten, in denen gewellte Membranstrukturen und gitterförmige Abstandhalter untersucht wurden, werden in dieser Arbeit Membranprofile mit spitzen und trapezförmigen Nuten ausgewählt. Die auf beiden Membranseiten quer zu einander angeordneten Nuten bilden im direkten Kontakt mit einer gleichartigen Membran ca. 1 mm tiefe Kanäle für den Wassertransport durch den Membranstapel.

Die konventionellen Herstellungsmethoden von sogenannten homogenen Ionenaustauschermembranen beruhen auf dem Ausziehen einer Polymerlösung auf einer geeigneten Unterlage und dem nachfolgendem Verdampfen des Lösungsmittels. Die Herstellung von sogenannten heterogenen Membranen erfolgt durch Erhitzen und Pressen oder Extrusion eines Gemisches aus Ionenaustauscherharz und Binderpolymer. Beide Verfahren eignen sich auch für die Herstellung von profilierten Membranen. Dabei können mit dem Verfahren, das auf dem Ausziehen und Verdampfen einer Polymerlösung beruht, nur einseitig profilierte Membranen hergestellt werden. Mit dem Extrusionsverfahren dagegen lassen sich auch beidseitig profilierte Membranen herstellen. Das Verfahren ist einfach und eignet sich auch für eine industrielle Produktion von beidseitig profilierten heterogenen Membranen.

Daher wurden die Membranen für die anschließenden Entsalzungsversuche durch Verpressen bzw. Extrusion von Mischungen aus gemahlenem Ionenaustauscherharz und einem Binderpolymer hergestellt. Tests wurden durchgeführt, um geeignete Polymermischungen auszuwählen, die die geforderten Eigenschaften der Membranen in Bezug auf Leitfähigkeit, Quellung und mechanische Stabilität gewährleisten. Neben den selbst hergestellten Membranen wurden auch kommerzielle heterogene Membranen mit Hilfe eines mit Riffelwalzen versehenen Kalanders beidseitig mit Profilen versehen.

Die profilierten Membranen wurden dann in einem Membranstapel eingesetzt, dessen Konstruktion eine Kontamination des Diluats, d.h. des Produktwassers, durch Co-Ionen aus dem Konzentrat weitestgehend verhinderte. Dies wurde dadurch erreicht, dass *a*) das Diluat aus den Diluatkammern, die neben den Elektrodenkammer angeordnet sind, nicht zusammen mit dem Diluat aus anderen Diluatkammern vermischt wird, sondern mit dem Konzentrat verworfen wird; und *b*) die Flussrichtung in den Konzentratkammern gegenläufig zur Flussrichtung in den Diluatkammern gewählt wurde. Außerdem wird ein Teil des Diluats zur Spülung der Konzentratkammern genutzt. Die Elektrodenkammern, die konventionelle Abstandshalter (Spacer) zwischen Elektrode und Membran besitzen, werden mit enthärtetem Leitungswasser gespült, um eine ausreichende Leitfähigkeit zu gewährleisten. Ein Membranpaket mit konventionellen nicht profilierten Membranen und ein Paket mit profilierten Membranen sind für vergleichende Versuche zusammengebaut.

In einer Versuchsreihe wurde das Paket mit profilierten Membranen mit deionisiertem Wasser in den Diluat- and Konzentratkammern getestet. Durch Bestimmung der Ionenkonzentration in beiden Kammern konnte die durch Ionentransport aus den Elektrodenkammern verursachte Kontamination des Produktwassers abgeschätzt werden. Es zeigt sich, dass eine nicht ausreichend hohe Membranselektivität eine relativ hohe Kontamination des Produktwassers durch Ionen aus den Elektrodenkammern zur Folge hat und damit die Möglichkeit zur Herstellung von Wasser mit einer sehr geringen Leitfähigkeit mit Hilfe der Elektrodialyse begrenzt ist.

Entsalzungsversuche mit enthärtetem Leitungswasser oder mit Umkehrosmose-Permeat als Rohlösung und gepressten heterogenen profilierten Membranen zeigen, dass mit höheren Stromdichten und geringerer Durchlaufgeschwindigkeit der Rohlösung die Leitfähigkeit im Diluat abnimmt. Bei Versuchen mit Umkehrosmose-Permeat strebt die Leitfähigkeit im Diluat mit Erhöhung der Stromdichten einem minimalen Grenzwert zu, der auch bei einer weiteren Stromerhöhung nicht unterschritten werden kann. Insgesamt haben die Versuche mit Umkehrosmose-Permeat als Rohlösung gezeigt, dass eine

Mindestleitfähigkeit im Diluat von ca. 2 µS/cm in einem einstufigen Prozess nicht unterschritten werden kann. Aber auch in einem mehrstufigen Entsalzungsversuch war es nicht möglich, die Diluatleitfähigkeit wesentlich unter einen Grenzwert von 1 µS/cm abzusenken.

Dass der Grenzwert der Diluatleitfähigkeit von 1-2 µS/cm nicht unterschritten werden konnte lässt sich auf zwei Phänomene zurückführen:

1) Die Permselektivität von heterogenen Ionenaustauschermembranen ist relativ gering.

2) Die Ionenaustauschkapazität und die spezifische Membranfläche von heterogenen Membranen sind niedriger als bei homogenen Membranen. Hinzu kommt, dass die Oberfläche von heterogenen Membranen zu einem großen Teil mit dem Binderpolymer bedeckt ist und daher nicht für den Austausch von Ionen aus dem Prozess-Wasser zur Verfügung steht. Auch die durch Wasserspaltung an der Oberfläche von heterogene profilierten Membranen entstehenden H^+- und OH^-- Ionen tragen nur wenig zum Austausch mit schwachdissoziierten Elektrolyten aus dem Wasser bei, sondern werden zum größten Teil in die Konzentratkammern weitergeleitet.

Durch die auf 1-2 µS/cm begrenzte Mindestleitfähigkeit des Diluats ist die Elektrodialyse mit heterogenen profilierten Membranen keine Alternative zur kontinuierlichen Elektrodeionisation, bei der das Diluat eine Leitfähigkeit von 0,055-0,2 µS/cm nicht überschreiten darf.

Bei einer Nutzung von homogenen profilierten Membranen mit höherer Ionenaustauschkapazität, deren Entwicklung in Zukunft durchaus möglich ist, kann die erreichbare Mindestleitfähigkeit allerdings noch erheblich erniedrigt werden.

Ein Vergleich der Entsalzungsleistungen einer konventionellen Elektrodialyse mit flachen Membranen und Gitter-Spacern mit einer Elektrodialyse mit profilierten Membranen aus gleichem Material und unter gleichen Testbedingungen zeigt jedoch, dass sich bei der Elektrodialyse mit profilierten Membranen eine erheblich bessere Entsalzung bei geringerem Energieverbrauch erzielen lässt. Ebenso haben die experimentellen Ergebnisse gezeigt, dass bei gleichem Entsalzungsgrad eine Elektrodialyse mit profilierten Membranen eine geringere Membranfläche benötigt und weniger Energie verbraucht als eine konventionelle Elektrodialyse mit flachen Membranen.

Auch die Versuche mit kommerziellen heterogenen Membranen, die mit Hilfe von Riffelwalzen profiliert wurden, bestätigen die mit den gepressten profilierten Membranen erreichten Entsalzungsgrade und den geringen Energieverbrauch.

Die technischen und wirtschaftlichen Vorteile der Elektrodialyse mit profilierten Membranen, die sich aus dem höheren Entsalzungsgrad und dem niedrigerem Energie- und Membranflächenbedarf ergeben, zeigen, dass die Elektrodialyse mit profilierten Membranen eine gute Alternative für die konventionelle Elektrodialyse mit flachen Membranen und Abstandshaltern ist. Profilierte Membranen erweitern die Einsatzmöglichkeiten der Elektrodialyse besonders der Herstellung von Industriewasser hoher Qualität ganz erheblich.

Im vierten Kapitel werden die Grundlagen der kontinuierlichen Elektrodeionisation beschrieben, ein Verfahren, das heute immer häufiger den konventionellen Ionenaustausch ersetzt und hauptsächlich zur Herstellung von Prozesswasser hoher Qualität eingesetzt wird. Dabei werden unterschiedliche Konzepte eingesetzt, die sich hauptsächlich die Verteilung der Kationen- und Anionenaustauscherharze in den Diluatkammern unterscheiden. Zu den am häufigsten eingesetzten Konzepten gehören die Füllungen der Diluatkammern mit Kationen- und Anionenaustauscherharz: a) In einem Mischbett; b) In einem mit aufeinander folgenden Schichten von Kationen- und Anionenaustauscher Bett; oder c) In getrennten Betten. Die Konzepte mit geschichtetem Bett oder getrennten Betten haben gegen über dem Konzept mit einem Mischbett den Vorteil, dass schwachdissoziierte Säuren besser entfernt werden wobei die Materialkosten niedriger sind.

Im fünften Kapitel wird die kontinuierliche Elektrodeionisation mit getrennten Kationen- und Anionenaustauscherbetten betrachtet, bei der die zur Regenerierung der Ionenaustauscher notwendigen H^+- und OH^--Ionen durch Wasserspaltung in einer Bipolarmembran zwischen den getrennten Betten erzeugt werden. Mit diesem verbesserten Verfahren konnte eine sehr gute Entfernung von schwachen Säuren erzielt werden. Allerdings kann es dabei häufig zu einem ein Durchtritt von Ionen aus dem Konzentrat in das Diluat, sodass es schwierig war, die gewünschte niedrige Leitfähigkeit im Diluat zu erreichen.

Um das durch die nicht vollständige Permselektivität der Membranen hervorgerufene Problem der Kontamination des Diluats bei der kontinuierlichen Elektrodeionisation mit getrennten Betten und Bipolarmembranen zu überwinden, wurden verschiedene Lösungsansätze untersucht. So wurde vorgeschlagen, eine mit Anionenaustauscherharz gefüllte Schutzkammer zwischen der Diluat- und Konzentratkammer einzuführen, oder aber die Konzentratkammer mit einem Anionenaustauscherharz füllen. Dabei sollte die Schutzkammer oder Konzentratkammer entweder mit dem Permeat der Umkehrosmose oder mit einem Teil des Diluats gespült werden.

Im sechsten Kapitel werden die experimentellen Untersuchungen der Konzepte der kontinuierlichen Elektrodeionisation mit getrennten Kammern und Bipolarmembranen mit und ohne Schutzkammer beschrieben. Dazu wurde eine Testanlage aufgebaut, in der beide Verfahren parallel unter exakt gleichen Bedingungen getestet werden konnten. Die Versuchsergebnisse haben gezeigt, dass sowohl mit dem Verfahren ohne Schutzkammer als auch dem mit Schutzkammer eine effektive Entfernung aller Elektrolyte erzielt werden kann. Allerdings durchläuft die Diluatleitfähigkeit mit Erhöhung der Stromdichte bei dem Verfahren ohne Schutzkammer ein Minimum. Dies kommt durch zwei gegenläufige Effekte zustande. Einmal werden durch die Erhöhung der Stromdichte in der Bipolarmembran mehr H^+- und OH^--Ionen erzeugt, die wiederum eine besseren Regenerierung der Ionenaustauscher zur Folge haben. Gleichzeitig erhöht sich aber mit Erhöhung der Stromdichte auch der Ionentransport aus dem Konzentrat in das Diluat. Auch eine Erhöhung des Drucks und Erniedrigung des Durchsatzes bei konstantem Druck in Konzentratkammer erniedrigt die Produktwasserqualität.

Das Konzept der kontinuierlichen Elektrodeionisation mit Bipolarmembranen, und getrennten mit Kationen- bzw. Anionenaustauschern gefüllten Diluatkammern, und einer mit Anionenaustauscher gefüllten Schutzkammer, wurde abschließend ausführlich untersucht. Dazu wurden bestimmte Kenngrößen, wie die Stromdichte, die Rohwasserleitfähigkeit, sowie die Durchflussgeschwindigkeit in der Schutz- und den Diluatkammern variiert. Auch der Einfluss unterschiedlicher Strömungsrichtungen in den Kammern wurde untersucht. Dabei zeigte sich, dass unterschiedliche Strömungsrichtungen in den Diluat-, Schutz- und Konzentratkammern einen erheblichen

Einfluss auf die Stromdichteverteilung und den Spannungsabfall haben und dadurch auch die Produktwasserqualität beeinflussen können.

Die Experimente haben gezeigt, dass sowohl Umkehrosmose-Permeat als auch Diluat zur Spülung der Schutzkammer oder Konzentratkammer benutzt werden können, wobei bei Verwendung des Diluats eine geringfügig bessere Leistung erzielt wird.

Mit den durchgeführten Untersuchungen wurde nachgewiesen, dass mit Hilfe der kontinuierlichen Elektrodeionisation mit Bipolarmembranen und Schutzkammer Reinstwasser mit einer Leitfähigkeit von ca. 18 MOhm·cm und einem Siliciumgehalt von unter 1 ppb erzeugt werden kann.

Damit konnten in der Arbeit sowohl für Elektrodialyse mit profilierten Membranen als auch für die kontinuierliche Elektrodeionisation vielversprechende Wege für Weiterentwicklung aufgezeigt werden.

1 Introduction

The demand for water of sufficient quality is rapidly increasing to meet the needs of a growing world population. Especially in the aride zones of the earth such as in the Middle East and the desert areas of Africa and Asia the traditional sources of potable water, i.e. surface or well water must be complemented by new technologies to generate potable water from sea or brackish water, or by recovering water from contaminated industrial effluents. In the northern hemisphere there is an increasing demand for water of high quality for industrial use and a recycling of industrial water becomes mandatory not only to save water but also to avoid contamination of the environment with toxic by-products of industrial processes.

The focus of this thesis is on the provision of water of low ion content for industrial purposes, as required for steam generation, for the production of fine chemicals or pharmaceuticals, for analytical purposes and for the electronics industry. This process water is usually made from tap-water in several purification steps.

1.1 Water desalination processes

In water purification different contaminants such as particulate organic and inorganic matter, microorganisms, salts, and eventually gases must be removed from raw water. The removal of particulate material by a sand bed or by membrane filtration is usually the first step. For the next steps a large number of processes such as adsorption, crystallization, distillation, ion-exchange and membrane based processes such as reverse osmosis or electrodialysis are available.

Distillation and crystallization

Desalination by distillation is based on the evaporation and re-condensation of water, and crystallization is based on the freezing of water and subsequent melting of ice. They rely upon the fact that saline impurities have negligibly low partition coefficients in the vapor or in the solid (ice) phase of water. Both processes are based on a phase change introduced by a temperature gradient. Regarding the energy consumption both processes are not very energy efficient and crystallization is also impaired by mass transfer problems and has very little relevance in water desalination. But distillation, using multi-effect evaporators is widely applied for desalination of raw water with high salt concentration such as sea water, in particular if waste heat obtained from power stations can be utilized for the evaporation process. It is however less often used for the purification of process water and will not be considered further.

Reverse osmosis

Reverse osmosis and also nanofiltration are pressure driven membrane separation processes which are based on the preferential transport of water through a dense separation layer of a membrane. After the development of suitable membranes and operation techniques reverse osmosis is the standard element in many modern water treatment plants. It is widely applied in brackish and sea water desalination. Compared to currently used alternative desalination technologies the desalination cost for reverse osmosis is the lowest for feed water salt concentrations from 3 000 ppm to 10 000 ppm, but due to comparatively low membrane costs and other factors reverse osmosis is usually applied for a much broader range of feed water concentrations [1]. Treating a feed water with above specified range of *TDS*, typical product water concentrations from reverse osmosis range from 50 ppm to 500 ppm of *TDS*.

Reverse osmosis is the preferred pre-purification step for high quality process water production, where it is used to purify tap-water. After reverse osmosis only a reduced amount of organic matters and salts will be present in the permeate, which has typically a *TDS* from 2 ppm to 50 ppm. If necessary, organics may be further reduced by adsorption, while the remaining salts have to be removed by methods which rely on electrostatic forces acting on the salt ions. Only these methods will be discussed in the following.

Ion-exchange

Ion-exchange is based on the exchange of ions between an ion-exchange resin and water. For the removal of salt traces, the water to be processed is usually flowing through a bed of ion-exchange beads, where the salt cations are exchanged with H^+-ions and the salt anions with OH^--ions, while H^+-ions and OH^--ions combine to neutral water. By ion-exchange deionized water with a very low level of residual ionic contaminants can be obtained. Since the ion-exchange capacity of the resin will be exhausted after some time, it has to be periodically regenerated with chemicals. To provide a sufficiently long period between regenerations, ion-exchange should be restricted to the treatment of feed water with an ionic content below 500 ppm [1]. Ion-exchange resins can also be processed as membranes, which have the ability to be either selective for cations or for anions. This allows to setting up electro-membrane processes where ions can be separated from the water by an externally applied electric field.

Electrodialysis

Electrodialysis is the most established electro-membrane process for water purification [8]. The process principle is shown in **Fig. 1.1**. The feed water is divided into two flows, directed through the diluate (DC) and the concentrate compartments (CC). The ion-exchange membranes and the applied electrical field are positioned in a way that both cations and anions are removed from the diluate compartments through the corresponding membranes into the concentrate compartments. The feed water is thus separated into a diluate (usually about 50 to 80% of the feed) and a concentrate. Conventional electrodialysis is well suited for treatment of feed water with salt concentrations of 500 ppm to 3000 ppm leading to about 10 ppm in the diluate [1] at relatively low process cost.

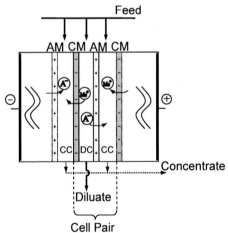

Fig. 1.1. Scheme of a simple electrodialysis unit. DC and CC are diluate and concentrate compartments; AM and CM are anion- and cation-exchange membranes respectively

Electrodeionization

Electrodeionization is a combination of electrodialysis and ion-exchange. This method is used as the final step to produce pure water or ultra-pure water with a conductivity in the range 0,055 µS/cm to 0,2 µS/cm, or with a content of some ions from a few ppb down to undetectable limits. Usually the feed water used for electrodeionization has conductivities lower than 50 µS/cm or a salt concentration lower than 25 ppm.

The further improvement of electrodialysis as well as of electrodeionization are the main goals of the present study.

1.2 The objectives of the thesis

The main objectives of the work described in this thesis are:

1) a development of a new electrodialysis concept with profiled membrane surfaces to reduce both the investment and energy costs for the production of deionized water for industrial applications;

2) a basic study of continuous electrodeionization and the development of a new process concept based on bipolar membranes with reduced costs for the production of ultra-pure water.

1.3 The structure of the thesis

At first the physicochemical fundamentals necessary for an understanding of electromembrane processes, including the properties of water and transport of ions in solutions and in ion-exchange materials will be summarized. The principle of ion-exchange, the structure of ion-exchange membranes and the main transport phenomena determining electro-membrane processes will be briefly reviewed.

Second, the development of new profiled membranes and their performance in electrodialysis under various experimental conditions will be investigated and the test results will be discussed.

In the third part of the thesis the principals and the currently applied concepts of electrodeionization technology will be reviewed. Based upon this discussion, the development of improved continuous electrodeionization concepts, based on bipolar membranes and separated cation- and anion-exchange resin beds, will be described and the results of the experimental investigation will be discussed.

2 Fundamentals of electro-membrane desalination processes

For a better understanding of the physicochemical phenomena occurring in electromembrane desalination processes some basic aspects of the structure and properties of water and of electrolyte solutions shall be summarized. Furthermore, electrochemical and structural properties of ion-exchange resins and membranes as well as transport phenomena of ions in solution and ion-exchange materials will be reviewed.

2.1 Electrolyte solutions in water

The general structure of water and aqueous solutions of electrolytes will be described and the thermodynamic equilibrium between different components as well as the transport of various components in a solution caused by different driving forces will be discussed.

2.1.1 Physicochemical properties of water

The structure of water molecules consisting of one oxygen and two hydrogen atoms is schematically shown in the **Fig. 2.1***a*. The electron-pairs of oxygen have a partially negative charge and the two hydrogen atoms have a partially positive charge creating a quadrupole. In a more simplified way water is usually considered being a dipole as shown in the **Fig. 2.1***b*.

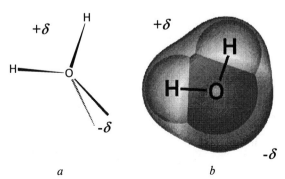

a   *b*

Fig. 2.1. Illustration of the water dipole: *a* – schematic representation of the molecule; *b* – distribution of the electron density [2]

Due to the dipole character of the water molecules, water has very special properties as summarized in Appendix 1. Water has a relatively low molecular weight and high concentration in the pure state or in most electrolyte solutions. Due to the small dimensions and the relatively high difference in polarities of oxygen and hydrogen, the dipole moment of water is large, making water a solvent with very high polarity and high affinity to charged or polarized species. The charge distribution of an isolated molecule is about $-0,7\,\bar{e}$ for the O-atom and an equal positive charge, shared between the H-atoms [3].

The geometry of water molecules calculated for an isolated molecule coincides with the experimentally determined values for gaseous water molecules, leading to 0,9578 Å for O-H length and 104,5° for the H-O-H angle [4,5]. In liquid water, due to the interactions between the molecules, the average values of the distance between O-H atoms and the H-O-H angle is higher and depends also on the presence of other components [6]. Commonly used molecular models apply O-H lengths of 0,957 Å to 1,02 Å and H-O-H angles of 104,4° to 109,5°.

Generally the hydrogen bond defines the association between an electronegative atom (e.g. N, O or F) and hydrogen, covalently bonded with another electronegative atom of the same or another molecule. With a few exceptions (mostly related to fluorine), the energies of hydrogen bonds are less than 20-25 kJ·mol^{-1}. Often compounds containing hydrogen show an intermolecular (or intramolecular, in case of long molecules) bonding, based on hydrogen bonds. The interaction between polarized water molecules forms hydrogen bonds which determine the different properties of water. In the liquid state the water molecules form numerous kinds of molecular clusters.

The mutual orientation of water molecules, electrostatic and electron interactions, allow H-atoms to be exchanged between water molecules by a protonation-deprotonation process, which is also called autoprotolysis of water:

$$2H_2O \rightleftarrows H_3O^+ + OH^-. \tag{2.1}$$

It results in the formation of a hydroxide anion OH$^-$, and a hydronium cation H$_3$O$^+$ (also called oxonium [8], oxidanium or hydroxonium ion). Both, acids and bases, catalyze the proton exchange of the autoprotolysis reaction.

The equilibrium constant (K_A) for the reaction described in Eq. (2.1) can be calculated by:

$$K_A = \frac{a_{H_3O^+} \cdot a_{OH^-}}{(a_{H_2O})^2},$$

(2.2)

where a is the activity of the corresponding species.

For simplification the hydronium ion H_3O^+ will be abbreviated as H^+ and the equilibrium described in Eq. (2.1) will be written as:

$$H_2O \rightleftharpoons H^+ + OH^-.$$

(2.3)

For dilute solutions, the activity of neutral water molecules can be considered as constant and equal to its concentration, which is 55,3 mol/L for the pure water at 25°C. This allows the use of the ionic product of water (K_W), also known as autoprotolysis constant, instead of the thermodynamic equilibrium constant:

$$K_W = a_{H^+} \cdot a_{OH^-}.$$

(2.4)

The ionic product of water at standard conditions is equal to $K_W = 1{,}008 \cdot 10^{-14}$ (mol/L)^2 and depends strongly on temperature.

The charged hydronium ion has an affinity to bind adjacent water molecules in different clusters [2,7], like $H_5O_2^+$ (Zundel cation), $H_9O_4^+$ (Eigen cation), or clusters combining larger number of water molecules with hydronium, e.g. $H_3O^+(H_2O)_6$ or $H_3O^+(H_2O)_{20}$.

2.1.2 Physicochemical properties of electrolyte solutions

The solutions are formed by dissolution of a matter in a solvent. In electrolyte solutions the solvent is generally water and the electrolytes are dissociated, forming charged mobile ions. The state of ions and molecules in electrolyte solutions and their transport under various conditions will be described below.

2.1.2.1 Dissolution of electrolytes

Due to the high affinity of water to polarized and charged species, the energy of hydration of such species is high. By dissolving electrolytes in water, the ionic or

8

covalent bond in the electrolyte will be broken due to the reaction with water, resulting in hydrated, positively and negatively charged ions in the solution. The hydration (also called aquation) of ions can be imagined as a sequence of consecutive steps from ion pairs to free hydrated ions as illustrated schematically in **Fig. 2.2**.

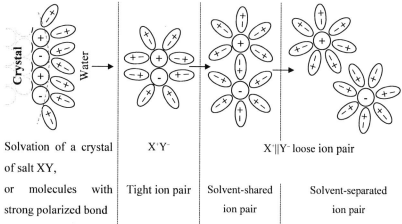

Solvation of a crystal of salt XY,	X^+Y^-	$X^+\|Y^-$ loose ion pair	
or molecules with strong polarized bond	Tight ion pair	Solvent-shared ion pair	Solvent-separated ion pair

Fig. 2.2. Schematic representation of subsequent steps during salvation of a crystalline salt

An ion pair is a pair of oppositely charged ions, held together by Coulomb attraction without formation of a covalent bond. The ion pairs exist in the solutions with high electrolyte concentrations, where they behave as electroneutral units with respect to conductivity, kinetic properties, osmotic behavior, etc. [8].

A formation of the second and following hydration shells increases the distance between the ions and rapidly reduces the electrostatic attraction between them. Thus the further hydration results in the formation of free hydrated ions in solution. The separation of a molecule or of an ion pair of the electrolyte into free ions is a dissociation reaction.

The free ions in solution are hydrated with water molecules, forming a hydration shell around the ion. The orientation of the molecules in the hydration shell breaks the hydrogen bonds with other water molecules and results in a net charge, distributed at this shell, which has the same sign as that of the ion in the centre. The charge at the first hydration shell can orient the next water molecules in the solution, forming a second hydration shell.

The number of water molecules in the hydration shell, also called the hydration number, will be reduced with increasing temperature and concentration of the solution. The hydration number increases with an increase of the ions charge density, which means that small ions tend to have a large hydration shell. Under the influence of an electrical potential gradient, ions move through the solution together with their hydration shell. Therefore small ions with a large hydration shell may be less mobile than large ions with a thinner hydration shell.

2.1.2.2 Dissociation of electrolytes

Dissociation can take place in pure substances such as molten salts and ionic liquids, or as a result of autoprotolysis in water. In the cases considered here, the dissociation is the result of the interaction with a solvent. Dissociation of an electrolyte in a solvent such as water leads to the formation of individual solvated cations and anions separated by solvent molecules. Due to the high hydration energy of charged species, the ionic bonds in crystals as well as strong polarized covalent bonds can be broken, which is typical for salts, bases and acids. Electrolytes, such as salts and most of the bases can exist partially as ionic pairs in solution at very high concentration (higher than 1 M), or dissociate completely in dilute solutions.

For covalently bound electrolytes, such as all acids and some bases the dissociation is considered as a reaction leading to ionization, and the common equation of thermodynamic equilibrium can be written as:

$$A_xB_y\,(aq) = xA^{z+} + yB^{z-}.$$

Thus, such electrolytes can be presented in solution in both forms: as ions and as neutral molecules, existing in equilibrium. The constant of equilibrium, on an example of a not completely dissociated acid, is:

$$K' = \frac{a_{H_3O^+} \cdot a_{A^-}}{a_{HA} \cdot a_{H_2O}}, \tag{2.5}$$

For dilute solutions $a_{H_2O} \approx const$; then the equilibrium constant is usually simplified as:

$$K = K' \cdot a_{H_2O} = \frac{a_{H_3O^+} \cdot a_{A^-}}{a_{HA}}. \tag{2.6}$$

The components present in solution mostly in non-dissociated form are considered as weakly dissociated electrolytes. One example of a typical weakly dissociated electrolyte in natural waters is silica, which can be present in different ionic forms, as well as in non-dissociated molecular forms, or as a polymer which is often referred to colloidal silica. The common silica present in form of mono- and bisilicate is generally referred to as reactive silica. The form of silica in solution depends strongly upon its concentration, the solution pH and upon the concentration of other components.

By the hydrolysis of some components the water molecules not only hydrolyze the ions during the dissociation of ionic or polarized covalent bonds, but also take part in the ionization reaction as donor of OH^-- or H^+-ions. The rest of water molecule, e.g. H^+- or OH^--ion is left as one product of this reaction and such electrolytes are called as Lewis acids and bases. Thus, for boric acid, as an example of Lewis acids:

$$H_3BO_3 + H_2O \rightleftharpoons B(OH)_4^- + H^+, \tag{2.7}$$

and for ammonia as an example of a Lewis base:

$$NH_3 + H_2O \rightleftharpoons NH_4^+ + OH^-. \tag{2.8}$$

2.1.2.3 The electro-chemical potential of ions in solution

Thermodynamically, the state of ions is characterized by their electrochemical potential, which is the partial molar Gibbs energy of an ion at the specified electric potential. Thus, the electro-chemical potential ($\tilde{\mu}_i$) of an ion i characterizes the state of this ion at defined conditions of pressure, temperature, activity and electric potential and is equal to:

$$\tilde{\mu}_i = \mu_i^\circ + V_{m_i}P + RT\ln a_i + z_i F\phi, \tag{2.9}$$

where μ° is the standard chemical potential, V_{m_i} - the partial molar volume, P – the pressure, a – the molar activity, T – the temperature, ϕ – the electric potential, z – the charge number, R – the gas constant, F – the Faraday constant; the subscript i refers to the specified ion.

The activity of the component is proportional to its concentration and given by:

$$a_i = \gamma_i C_i, \tag{2.10}$$

where γ_i is the activity coefficient and C_i is the molar concentration of the ion i.

The electro-chemical potential of each component inside a closed system in equilibrium is equal in every point of the system.

2.1.3   Transport phenomena in electrolytes

The main categories of transport processes in electrolytes are the hydrodynamic (also called convective) flow and the transport on molecular level. The hydrodynamic flux of the volume (J_V) is driven by the pressure gradient (∇P) and the flux of a component i transferred with the volume is:

$$J_i^V = J_V C_i = -C_i L_h \nabla P, \tag{2.11}$$

where L_h is the specific hydraulic permeability of the media, C_i – the molar concentration of the component i in the flowing volume.

The gradient $\nabla \tilde{\mu}_i$ of the electrochemical potential Eq. (2.9) is generally considered to be the driving force for the transport of a component on molecular level:

$$\nabla \tilde{\mu}_i = \nabla \left(V_{m_i} P \right) + \nabla \left(RT \ln a_i \right) + \nabla \left(z_i F \phi \right). \tag{2.12}$$

Assuming a linear dependency of flux J_i upon the electrochemical potential gradient $\nabla \tilde{\mu}_i$ (Eq. (2.12)) with $\dfrac{D_i C_i}{RT}$ as proportionality coefficient, the flux of a component with a constant charge is given by:

$$-J_i = \frac{D_i C_i}{RT} \nabla \left(P V_{m_i} \right) + \frac{D_i C_i}{T} \nabla \left(T \ln a_i \right) + \frac{z_i F D_i C_i}{RT} \nabla \phi, \tag{2.13}$$

where J is the vector of flux, D – diffusion coefficient, C – molar concentration, T – temperature, P – pressure, V_m – partial molar volume, γ – activity coefficient, z – ion charge number, ϕ – electric potential, R – universal gas constant.

The diffusion coefficient depends on the concentration of all components and on the temperature in the locally considered volume. In case of significantly different concentrations and temperatures in different points of the system the complex functional dependence including three variables $\left(\dfrac{D_i C_i}{T} \right)$ should be considered by solving Eq. (2.13).

The transport described by the first term of Eq. (2.13) with a pressure gradient as a driving force is called pressure diffusion. It has some importance in solids. In liquids however, the pressure gradient acts on all species in the same way, leading to a convective movement. For solutions this term can therefore be substituted by the convective flux of Eq. (2.11).

The second term in Eq. (2.13) refers to a flux due to temperature and concentration gradients, called diffusion. Since no significant temperature gradient is present in typical electromembrane desalination processes, the thermodiffusion is negligibly low and the second term can be simplified to molecular diffusion only ($D_i \nabla C_i$).

Finally, the third term of Eq. (2.13) presents the transport driven by the gradient of an electric potential and is called electromigration, or just migration. This transport is relevant only for charged species, e.g. ions and is the key transport mode in electromembrane processes.

Assuming isothermal conditions Eq. (2.13) gives:

$$-\boldsymbol{J}_i = \frac{D_i C_i}{RT} \nabla\left(PV_{m_i}\right) + D_i C_i \nabla\left(\ln a_i\right) + \frac{z_i F D_i C_i}{RT} \nabla\phi. \tag{2.14}$$

Introducing Eq. (2.10) into Eq. (2.14) transforms it into:

$$-\boldsymbol{J}_i = \frac{D_i C_i}{RT} \nabla\left(PV_{m_i}\right) + D_i C_i \nabla\ln\gamma_i + D_i \nabla C_i + \frac{z_i F D_i C_i}{RT} \nabla\phi. \tag{2.15}$$

If the concentration of the component does not differ significantly it can be assumed that D_i and V_{m_i} are constant so the gradient in the first term of Eqs. (2.14) and (2.15) will be modified into $\nabla\left(PV_{m_i}\right) = V_{m_i} \nabla P$.

As will be considered further (see Chapter 2.1.3.1) for the limit of zero concentration the relation between diffusion coefficient and the electric mobility of an ion is described by Nernst-Einstein equation:

$$D_i = u_i \frac{RT}{|z_i|F}. \tag{2.16}$$

Assuming that the activity coefficient γ_i in the concentration range of interest is constant, i.e. $\nabla\ln\gamma_i = 0$, the activity in the second term of Eq. (2.14) can be replaced by concentration and the relation between the D_i and u_i shown in Eq. (2.16) can be applied in the third term of Eq. (2.14). Introducing these assumptions into Eq. (2.14) leads to:

$$-\boldsymbol{J}_i = D_i \frac{C_i V_{m_i}}{RT} \nabla P + D_i \nabla C_i + \frac{z_i}{|z_i|} u_i C_i \nabla\phi. \tag{2.17}$$

Eq. (2.17), called Nernst-Planck equation, combines the fluxes of diffusion and migration and is often used in electromembrane processes to describe the transport of ions and neutral molecules on molecular level. The first term of Eq. (2.17), describing the pressure diffusion, is negligible in most of the cases and moreover could not be distinguished from the hydrodynamic flow. If the volume transport of the solution (Eq. (2.11)) in form of hydrodynamic flow, including convective streams, turbulent vortexes, etc. occurs, then the term $C_i J_V$ should be added to Eq. (2.17). This results in the extended Nernst-Planck equation of the form:

$$- J_i = D_i \nabla C_i + \frac{z_i}{|z_i|} u_i C_i \nabla \phi + C_i L_h \nabla P \,, \qquad (2.18)$$

where L_h is the hydrodynamic permeability of a medium where volume flow takes place.

If convection is negligible, the acting driving forces are due to concentration gradients and due to electrical potential gradients, which may either be imposed externally or created internally by separation of ionic charges.

2.1.3.1 Electromigration of electrolytes in solution

Diffusion is the transport of mobile species, neutral, as well as charged, under a gradient of its concentration and the diffusive flux in steady-state is proportional to the concentration gradient, with diffusion coefficient as proportionality factor. It is present as the first term in (2.18) and for one-dimensional case is:

$$- J_i = D_i \frac{dC_i}{dx} \,, \qquad (2.19)$$

where D_i and C_i are the diffusion coefficient and the concentration of i, and x is the coordinate.

Once an electric field is applied to an electrolyte solution the ions start to be driven by electric force and the positive ions, e.g. cations are migrating in the direction of electric field and negative ions, e.g. anions – in the opposite direction. By definition the electric field is the negative gradient of electric potential and acts here as a driving force for the migration. Compared to diffusion migration represents a more "active" transport. It affects only charged species and results in the transport of opposite charges in opposite directions. Pure migration can be observed in absence of the gradients of concentration, temperature and pressure.

If a migrative transport is in steady-state, than the electric force accelerating the ions is balanced by friction in viscous media and the ions are moving with constant velocity and the electric mobility is used to characterize the mobility of an ion in electric field. The electric mobility of ion u_i is defined as an average linear velocity of the ion i in a constant electric field of strength of 1 V/m.

Like the diffusion coefficient, ionic mobility is related to the dimensions of ion, but also is directly proportional to the charge number of ion. Thus, considering two hypothetic ions with the same effective diameter, one with charge number $z_i = 1$ and another one with $z_i = 2$, both of them will have the same diffusion coefficient, but the mobility for the later one is doubled.

Transport of H^+- and OH^--ions in water differs from the transport of hydrated mineral ions because of specific Grotthuss mechanism for charge transfer in water. In this mechanism the charge of H^+-ion is transferred through the chain of water molecules bonded with H-bonds inside a cluster, which significantly increases its transfer rate [7,9] making H^+-ion to the most mobile ion (see **Table 2.3**). Assuming that the effective diameter of hydrated H^+-ion is similar to K^+-ion, and comparing the mobility of both ions one can see that the mobility of H^+-ion is around five times higher. Thus, it can be suggested that 1/5 of the H^+-ions mobility is due to the usual migration similar to K^+-ions and 4/5 due to the Grotthuss mechanism. Similarly, the Grotthuss mechanism is valid for transport of OH^--ions in water.

Like the diffusion coefficient, the mobility of ions increases with temperature and decreases with concentration. At given temperature the highest mobility is reached in an infinitely diluted solution, because the distance between ions is infinitely long and there is no interaction between ions. With increase of concentration the distance between ions decreases and the forces of electrostatic interaction between them are no more negligible, influencing the mobility of ions in solution. The dependence of ionic mobility upon the concentration in dilute solutions is characterized by Debye-Hückel-Onsager theory, which is described in detail elsewhere [10,11,12].

The average scalar speed of an ion at given electric potential gradient is $u_i \cdot \left| \dfrac{d\phi}{dx} \right|$ and the migrative flux in the solution with molar concentration C_i for cations is:

$$J_+ = - u_+ \cdot C_+ \frac{d\phi}{dx} , \qquad (2.20)$$

and for anions is:

$$J_- = u_- \cdot C_- \frac{d\phi}{dx} . \qquad (2.21)$$

Equations (2.20) and (2.21) can be summarized in the common equation:

$$J_i = -\frac{z_i}{|z_i|} \cdot u_i \cdot C_i \frac{d\phi}{dx}. \tag{2.22}$$

Assuming the solution, where the sum of fluxes via diffusion (Eq. (2.19)) and migration (Eq. (2.22)) is equal to zero, e.g. the system is in equilibrium, the following correlation between D_i and u_i can be obtained [12]:

$$D_i = u_i \frac{RT}{|z_i|F}\left(1 + \frac{d\ln\gamma_i}{d\ln C_i}\right), \tag{2.23}$$

where γ_i and C_i are the molar activity coefficient and the concentration of i in solution.

For the infinitely diluted solutions, where the changes of interaction between the ion and its ionic sphere with concentration during the diffusion are negligible, i.e. $\left(\frac{d\ln\gamma_i}{d\ln C_i} \approx 0\right)$ Eq. (2.23) can be simplified into the Nernst-Einstein relation in the form of Eq. (2.16).

According to Eq. (2.22) the values of fluxes for ions with different concentrations and mobilities can be different. The ratio between the charge conducted by an ion Fz_iJ_i and the total charge transfer $\Sigma(Fz_iJ_i)$ characterizes the role of the given ion in the charge transport process and is called transport number of the ion (t_i):

$$t_i = \frac{z_i J_i}{\sum_i z_i J_i} = \frac{|z_i|u_iC_i}{\sum_i |z_i|u_iC_i}, \tag{2.24}$$

where z is the ion charge number, C - the molar concentration, u - the mobility and J - the flux of an ion i.

2.1.3.2 Electrical conductivity of solution

Ions moving in an electric field cause the net flux of electric charges, and thus conduct an electric current. The current density (i) is the sum the charge fluxes of all ions, which are given according to Eq. (2.22) by:

$$i = \frac{I}{A} = F\sum_i z_i J_i = -\frac{d\phi}{dx}F\sum_i |z_i| \cdot u_i \cdot C_i , \tag{2.25}$$

where A is the area conducting the current I, assuming parallel lines of electric field and ion fluxes.

Proportionality coefficient between current density i and electric potential gradient $\left(-\dfrac{d\phi}{dx}\right)$ in Eq. (2.25) characterizes the ability of solution to conduct electric current and is called as specific electrical conductivity (κ), mentioned further just as conductivity:

$$\kappa = F\sum_i |z_i| \cdot u_i \cdot C_i = \frac{i}{-\operatorname{grad}\phi}. \qquad (2.26)$$

While the mobility is used to characterize an ion at specified conditions, the ionic transport number is used to characterize a contribution of an ion in the total conductivity of electrolyte solution containing different ions.

Taking into account Eqs. (2.24) to (2.26), the transport number of an ion i in the Eq. (2.24) can be rewritten as a part of current conducted by ion i:

$$t_i = \frac{Fz_i J_i}{i} = \frac{|z_i| u_i C_i}{\kappa}. \qquad (2.27)$$

For the homogeneous electrolyte solution between parallel electrodes with surfaces area A and distance Δx between them the integration of Eq. (2.26) gives:

$$\kappa = \frac{\Delta x}{A}\frac{I}{\Delta\phi}. \qquad (2.28)$$

By measuring the potential drop and the current the conductivity of solution can be determined. Determining the conductivity by Eq. (2.28) assumes that no concentration changes in solution (for example, due to electrode reactions) occur. Thus, the practical measurements of conductivity are usually done at alternating current, with very low amplitude, to prevent a net transport of ions.

Furthermore, the determination of the conductivity of a solution according to Eq. (2.28) by measuring the potential drop and the current is affected by the electric field since it is difficult to keep only parallel lines of electric field between the electrodes in the cells used for practical measurements of the conductivity. If the electrodes or field lines between them are not parallel the effective distance between the electrodes is difficult to calculate. Thus, the geometrical parameters of the conductivity cell A and Δx in the Eq. (2.28) are combined in one proportionality coefficient between conductivity (κ) and conductance ($G = 1/R = I/\Delta\phi$), which is referred as the cell constant K_{cell}:

$$\kappa = \frac{K_{cell}}{R}, \qquad (2.29)$$

where R is the electric resistance.

The constant of the conductivity cell is determined by measuring conductance or resistance of the cell with solution of known conductivity.

Due to the direct proportionality between concentration and conductivity in Eq. (2.26), and because of the conductivity measurements in electrolyte solutions are very simple, the value of conductivity is often used to characterize the ionic content in solutions, for example to estimate a content of dissolved salts in water.

2.1.3.3 Dependence of conductivity upon concentration of impurities in water

Pure water, without any impurities, contains only two ions – H^+- and OH^--ions, which are the result of the autoprotolysis of water (Eq. (2.3)) and which are the cause for the (ultra-) pure water conductivity of $0,0550\ \mu S/cm$ at $25°C$. If additional ions are added, the conductivity will increase according to Eq. (2.26). In low conductivity waters, where the concentration of electrolyte is comparable with the concentration of H^+- and OH^--ions in pure water, water dissociation must be taken into account for the calculation of the conductivity.

It is shown in literature [13,14] that water with an extremely small amount of bases, such as NaOH in a concentration of less than $3,8 \cdot 10^{-8}\ M$, can have a specific conductivity slightly lower than the theoretical conductivity of pure water, as illustrated in the **Fig. 2.3**. Here the conductivities of HCl, NaOH and NaCl as a function of the Na^+- or Cl^--ions concentration in water are shown. The calculated $\kappa_D - C_{NaOH}$ dependence, starting from the value for pure water at zero NaOH concentration shows first a decrease of κ with C_{NaOH} and after reaching the minimum value of $\kappa = 0,0548\ \mu S/cm$ ($\rho = 18,25\ MOhm \cdot cm$) the conductivity increases again with increasing C_{NaOH}.

This phenomenon is related to the significantly higher mobility of H^+-ions compared to all other ions in the solution. The water can be considered as a weak acid, and addition of a base (NaOH) results in the decrease of the H^+-ion concentration, similar to the titration of an acid. If the amount of the added base is very small (only a part of the H^+-ion concentration in pure water), the replacement of most mobile H^+-cations with the slower cations of the base (Na^+) leads to the conductivity decrease.

18

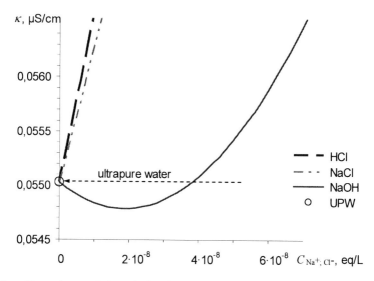

Fig. 2.3. Dependence of the calculated water conductivity upon the amount of trace level electrolytes at 25°C

The conductivity of pure water is very sensitive to the presence of different ionic contaminants. Pure water can be easily contaminated by contacting other substances, like materials of tanks, bottles, pipelines, air, etc. In practice, usually polymers of special grade are chosen for the transport of pure water to prevent its contamination.

In the contact with air gases will be dissolved in water. Most of them (N_2, O_2, Ar) have no influence on the water conductivity. But CO_2 reacts with water, forming carbonic acid, which is dissociating in two stages.

$$CO_2 + H_2O \rightleftharpoons H_2CO_3, \qquad (2.30)$$

$$H_2CO_3 \rightleftharpoons H^+ + HCO_3^-, \qquad (2.31)$$

$$HCO_3^- \rightleftharpoons H^+ + CO_3^{2-}. \qquad (2.32)$$

Ions produced in the reactions described in Eqs (2.31) and (2.32) readily affect the conductivity of water. Thus, water saturated with pure CO_2 has a conductivity around 50 µS/cm. Water equilibrated with air has a conductivity around of 0,7 - 1 µS/cm, caused by the presence of certain concentration of CO_2 in air, as shown in the **Table 2.1**.

Table 2.1

Influence on dissolved carbon dioxide on the specific water conductivity [15]

	κ, $\mu S/cm$
Ultrapure water	0,055
Water in equilibrium with air (0,5 mg/L of dissolved CO_2)	ca. 0,7
Water with 3 mg/L dissolved CO_2	ca. 2
Water with 6 mg/L dissolved CO_2	ca. 6

It is obvious that only dissociated electrolytes contribute to the electric conductivity. For example, compounds of silica and boron, present in natural waters, have very low dissociation constants and it is very difficult to detect the content of these compounds by conductivity measurements.

2.1.3.4 Dependence of conductivity upon temperature

An increase of temperature leads to a reduction in the average diameter of hydrated ions and in the viscosity of water. Both effects result in an increase of ions mobility and solution conductivity. For diluted solutions of salts the conductivity increase is usually ca. 2% per Kelvin. With rising temperature the dissociation degree of weak electrolytes as well as of the water molecules also increases, which additionally contributes to the conductivity increase of aqueous solutions.

The effect of temperature on the conductivity of many electrolyte solutions is known and temperature coefficients are listed in the corresponding literature. For natural waters, usually a non-linear dependence of κ over T is applicable and used in standard conductivity meters for temperature compensation. The conductivity is usually displayed for a specified temperature (usually 25°C).

Microprocessor type conductivity meters are available for measurements in water with conductivity lower than 1 $\mu S/cm$, considering both the non-linear temperature dependence of the ion mobilities and the ionic product of water for the temperature compensation. An example of temperature compensation in such devices is illustrated in **Fig. 2.4**, where the contribution of water (H^+ and OH^-) and salt to the total conductivity is taken into account.

20

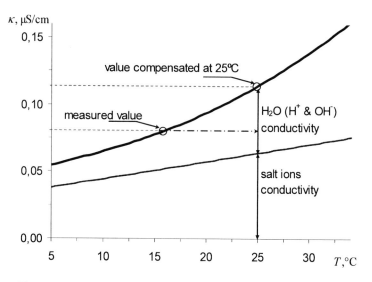

Fig. 2.4. Temperature dependence of conductivity in an example for water with neutral salt impurities and a measured overall conductivity of 0,114 µS/cm (fat line). A non-linear temperature dependence due to the salt is shown by the thin line [16]

It should be noted, that for water with basic or acidic impurities or with a significant content of low dissociated electrolytes this temperature compensation may result in some deviations from the real values. Weakly dissociated electrolytes have significantly lower influence on the water conductivity compared to strongly dissociated ones. At the same time they show a higher increase in conductivity with temperature because of an additional increase in the dissociation degree, compared to strongly dissociated electrolytes, where only the ion mobilities will be increased.

2.1.4 Gradation of water quality

There are different classifications of water based on the source of the water (natural water, waste water, etc.), or on its application (drinking water, process water, etc.) and on its content of salts or other impurities (pure water, salt water, etc.). In the following, only the salt content will be considered.

The parameter often used to specify the content of mineral ions in natural waters is the concentration of total dissolved solids (*TDS*) expressed in ppm (mg/L). For purified water with very low salt concentration the water quality is also expressed by its electrical conductivity or resistivity at 25°C.

For the characterization of the salt content in irrigation, drainage and ground waters the term "salinity" is used as total salt concentration expressed in mmol/L or mg/L. The same term "salinity" is used also in oceanology, and is defined as ratio between conductivity of specified water and a standard KCl solution. Also the "practical index of salinity" expressed as conductivity is often used in practice. For simplicity the different units of concentration and conductivity can be converted into each other, such as: 1 mS/cm = 10 mmol/L dissociated electrolytes = 700 mg/L *TDS* [17].

Natural waters contain mainly Na^+-, Ca^{2+}-, Mg^{2+}-, K^+-, HCO_3^--, SO_4^{2-}- and Cl^- -ions, and, depending on concentration, are separated in fresh waters, saline waters and brine. The *TDS*-range related to different kinds of natural water is not the same in different literature sources and depends on specifications in different countries [18,19]. The approximate *TDS* and κ values of waters are shown in the **Table 2.2**. Some particular classifications are summarized in the Appendix 2.

Table 2.2

Salt concentration in different classes of water

Water class	*TDS*, mg/L	κ, µS/cm
Fresh water	< 500	< 700
Saline (or salt) water	500 – 50 000	700 – 50 000
Brine	> 50 000	> 50 000

Sea water has usually a *TDS* value of about 35 000 mg/L and brackish water normally refers to water with a *TDS* value between 1 000 to 10 000 mg/L. These salt concentrations are appreciably higher than in drinking water, which has usually lower than 500 - 700 mg/L of *TDS*[*]. Also the water for irrigation should have a *TDS* concentration of less than 1200 ppm[**].

---

[*] in some countries up to 1000 mg/L TDS in drinking water is allowed [18]

[**] range allowed by FAO is 450 - 2000 ppm of TDS for irrigation [19]

The quality of high-purity water for boiler feed is mostly specified by a conductivity below 0,10 µS/cm and the sodium and silica concentrations to be < 10 ppb. The silica concentration in this application is critical to prevent precipitations in pipes or on turbines. Here silica may precipitate on the blades as a glassy deposit which reduces the turbine efficiency. Both types of silica, colloidal and reactive, can cause this problem since colloidal silica will volatilize under high temperature.

Also in the semiconductor industry the silica content is strictly limited, because it influences the properties of semiconductors by causing water spots on the wafers surface. Actually the semiconductor industry has the highest requirements to the water quality. The process water used in the electronic and semiconductor industry is called ultra-pure water. The specifications to ultra-pure water are given in the ASTM standard [20] and in International Technology Roadmap for Semiconductors [222]. It is generally defined by a resistivity of 18,2 MOhm·cm, concentration of metal ions below 1 ppt, inorganic anions and ammonia below 50 ppt, and total organic carbon below 1 ppb (see Appendix 3).

2.2 Ion-exchange

Mobile ions are not only present in liquid electrolyte solutions and melts, but they also exist in certain solids, such as polymeric electrolytes, which are most relevant for water treatment applications. In these materials ions of one charge are mobile and ions of the opposite charge are fixed. When the mobile ions from a solid electrolyte are in contact with an electrolyte solution, they can move to the solution and be replaced by the ions of the same charge from the solution. Such a phenomenon is called ion-exchange and such solid electrolytes are referred to as ion-exchangers. Ion-exchangers contain functional groups fixed to the solid polymer matrix which are either ionized or capable of dissociation into fixed ions and mobile counter-ions.

Most widely used ion-exchangers in practical applications are granulated synthetic polymeric resins. They are applied in large quantities in water softeners and deionization systems, but they are also used for the removal of specific ions such as radioactive ions, heavy metal ions, NO_3^-- and NH_4^+-ions, etc., as well as in some special applications in analytical methods or as reaction catalysts.

2.2.1   Structure and synthesis of ion-exchange resins

Depending on the charge of ion-exchange groups, ion-exchange resins are categorized in cation-exchange resins (CR), anion-exchange resins (AR) and ampholytes. Ion-exchange resins are produced in form of powder (particle size under 0,1 mm), beads (0,3 mm – 2 mm), fibers, sheets or blocks. By structure ion-exchange resins can be divided into gel-type, mesoporous and macroporous resins. By the dissociation degree of the fixed groups one can distinguish between strongly or weakly dissociated ion-exchange resins.

Different types of polymer backbones and fixed groups can build the solid structure of an ion-exchange resin. Common fixed ionic groups in cation-exchange resins are: $-SO_3^-$, $-PO_3^{2-}$, $-COO^-$; and in anion-exchange resins: $-N(CH_3)_3^+$, $-NH(CH_3)_2^+$, $-NH_2(CH_3)^+$.

Most ion-exchange resins in water treatment applications are in granulated form and have a backbone of polystyrene, cross-linked by divinylbenzene (DVB). The first step in the synthesis of such ion-exchange resins is usually the emulsion polymerization of the mixture from styrene and DVB with additives in water. Polymerization results in hard polydisperse beads of a *co*-polymer with a cross-linked structure, which is schematically shown in the **Fig. 2.5**. Resins based on cross-linked polyacryl or other kinds of polymers are also known and used for special purposes.

Spherical beads of resin produced by emulsion polymerization have a broad distribution of particle size. For the continuous production of beads with a narrow particle size distribution, special techniques, such as injection of monomers into hot-water [21,22] or the vibratory excitation of a laminar jet of a reactive monomer mixture into the liquid or gas phase [23,24] are used.

In the next synthesis step fixed groups will be covalently bounded to the polymer backbone. The reaction of polystyrene-*co*-divinylbenzene with hot sulfuric acid or with chlorosulfonic acid results in the sulfonation of almost all aromatic rings. The sulfonic groups will dissociate in water and form anions, which are covalently bound to the polymer ("fixed ions") and mobile cations ("counter-ions").

The synthesis of anion-exchange polymers leads to fixed ions with positive charges, which have anions as mobile counter-ions. Usually the synthesis on an anion resin includes a chloromethylation process followed by amination.

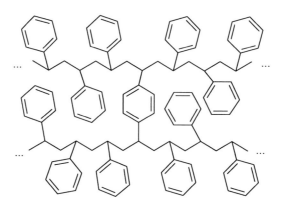

Fig. 2.5. Chemical structure of cross-linked polystyrene-*co*-divinylbenzene

The character of the polymer backbone and the type of the fixed ion-exchange group define the properties of the ion-exchange resin. Especially important parameters for ion-exchangers are the degree of cross-linking and the concentration of fixed groups.

It is known that in general polymeric anion-exchangers with OH⁻-ions and cation-exchangers with H^+-ions as mobile counter-ions have lower thermal and chemical stability than resins with salt ions as mobile counter-ions. Usually a polymeric cation-exchanger has higher thermal stability than an anion-exchanger. The data of thermogravimetry measurements of commercial anion-exchange resins (AR) with quaternary ammonium fixed groups and OH⁻-ion as counter-ion, and cation-exchange resins (CR) with sulfonic acid fixed groups and H^+-ion as counter-ion are shown in **Fig. 2.6**.

The curves in **Fig. 2.6** indicate the small decrease of mass for all samples between 50°C and 200°C which is due to evaporation of residual water from the samples. At higher temperatures the splitting of the quaternary ammonium groups takes place between 220°C and 250°C for the samples of ARs with a following destruction of the polymer backbone. The thermal destruction of CRs becomes visible only at temperatures higher than 350°C. The degree of destruction strongly depends on the time exposed to elevated temperature and on the presence of oxidative agents, such as O_2.

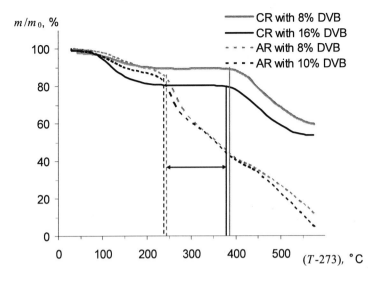

Fig. 2.6. Thermogravimetric mass loss in oxygen atmosphere of powdered samples of cation-exchange resins Dowex C400 (8% DVB) and Diaion SK116 (16% DVB), as well as anion-exchange resins Amberjet 4400 (8% DVB) and Diaion SA10A (10% DVB) as function of temperature

As mentioned above, the fixed groups in ion-exchange resins will by hydrated and dissociate in contact with water. Due to the cross-linking the polymer stays insoluble and the charged fixed ions remain bond to the polymer backbone. The hydrated counter-ions become more mobile but are attracted by the Coulomb force to fixed ions and the whole bead remains electroneutral. Due to the presence of charged groups ion-exchange resins are very hydrophilic and show high water uptake. Swelling stops when the difference of osmotic pressure within the resin and in the surrounding solution is counterbalanced by the elastic forces of the stretched polymer network. The degree of swelling therefore depends strongly on the degree of cross-linking as well as on the concentration and charge number of the ions present in the resin. The water content in the swollen resin is typically $(50 \pm 20)\%$ for commercial polystyrene based resins. The uptake of water by swelling results in the increase of the bead diameter. Usually, the volume of beads increases by a factor of 1,5-3 corresponding to the diameter increased by a factor 1,1-1,4 [25] depending on cross-linking degree and other parameters.

The absorbed water is located in the hydrated shells of the ions and in hydrophilic clusters built by fixed ions and counter-ions. The dimensions of such clusters are defined by the average distance between the cross-linking bridges of DVB and are usually in the range of 1 - 4 nm. Thus, if no pores are built during the synthesis the ion exchange-resins swollen in water have a typical gel structure [26].

Also ion-exchange resins with a specific porosity are produced for special applications. The pores in such resins are introduced during the synthesis using cross-linking agents with long molecular chains or additives which leave pores after polymerization.

Ion-exchange materials made in form of fibers or porous blocks are also known. Fibrous materials can be produced as non-woven fibers or woven spacers [27]. Today, some of such ion-exchange textiles are also commercially available and tested in medical applications [28] and in catalysis [29,30]. In contrary to resin beads, ion-exchange fibers and porous blocks are usually produced by grafting of fabrics or porous blocks made of a conventional polymer, like polypropylene, followed by chemical treatment resulting in introduction of fixed groups and cross-linking. Most commonly, irradiation grafting is used and such produced ion-exchange materials are named as "grafted". Due to restricted availability and relatively high prices for ion-exchange textiles, their usage in electromembrane processes is currently limited.

2.2.2   Donnan potential, osmotic pressure and co-ion exclusion

If a dry ion-exchange resin bead is put in pure water, water molecules will start to penetrate the polymer matrix causing a swelling of the bead. Now the counter-ions will dissociate from the ions fixed at the polymeric network and will be dissolved inside the gel. Because of the concentration gradient, the counter-ions also tend to diffuse from the resin into the outside water. In doing so, a charge difference develops. The resulting potential, called the *Donnan potential*, prevents counter-ions from leaving the pellet. However, water molecules from the outside continue to diffuse into the polymer pores due to the water concentration (or osmotic pressure) gradient. Swelling increases until the osmotic pressure is compensated by the pressure built by the stretched polymeric network. If instead of pure water a salt solution is at the outside, counter-ions may exchange between the resin bead and the outside solution due to their concentration gradients. But co-ions are prevented from entering the resin because of the Donnan potential.

These complex interactions can be well described by the transport equations developed in Chapter 2.1.3 and lead to an equilibrium if the electrochemical potential of each component is the same inside and outside of the resin, i.e. $\nabla\tilde{\mu}_i = 0$. This is possible if the three terms determining the electrochemical potential in Eq. (2.12) compensate each other.

Considering the water, a higher activity of water in the outside solution compared to the resin drives water into the resin and increases the osmotic pressure inside the polymer matrix, until the term with the activity gradient in Eq. (2.12) is equilibrated by the term with pressure and partial molar volume.

Considering mobile ions, an equilibrium concerning the pressure, the concentration and the electrochemical potential term has to be reached. The osmotic pressure of the water inside the resin is also acting on the ions in the resin. Since counter-ions are present in high concentration inside the resin, they diffuse out into the solution by activity and pressure gradients (see Eq. (2.13)). This causes an electrical potential gradient at the *resin/solution* interface, which compensates the other driving forces of the counter-ions. At the same time the mobile co-ions which have a higher concentration in the solution than in the resin, are hindered in penetrating into the resin by the above mentioned potential gradient.

The potential drop, due to the Donnan potential, located at the *resin/solution* interface, plays a decisive role in ion-exchange since in diluted solutions it excludes co-ions from entering the resin, the so-called "Donnan exclusion".

In general, the Donnan potential created by counter-ions can be derived from the Eq. (2.12) as:

$$\nabla\phi_{Don} = \frac{1}{Fz_i}\left(RT\ln\left(\frac{a_i}{\bar{a}_i}\right) + V_{m_i}\left(P - \bar{P}\right) \right) , \qquad (2.33)$$

where $\bar{a}_i$ and a_i are the activities of counter-ions in the resin and in the solution; $\bar{P}$ and P are the pressures inside the resin and the solution.

At low electrolyte concentrations the counter-ion activity and activity coefficients inside the resin are usually assumed to be constant and the concentration of counter-ions inside the resin is assumed to be equal to the concentration of fixed ions, the so-called *ion-exchange capacity*.

The Donnan potential therefore depends strongly on the ion-exchange capacity of the resin and the ion concentration of the solution. Also the charge and the type of ion-exchange groups and the counter-ions as well as the degree of cross-linking of the resin strongly influence the value of the Donnan potential.

Donnan potential impedes co-ion penetration, but does not affect non-dissociated molecules. This means that electrolytes present in non-dissociated form or as ionic pairs can be more readily absorbed by the resin than the charged co-ions, since they are only hindered by the osmotic pressure difference, not by Donnan-exclusion.

2.2.3 Ion-exchange selectivity

Counter-ions can move inside the gel structure of an ion-exchange resin and can also be exchanged with ions from the outside solution. Thus, if the counter-ion in the resin is not the same as in the surrounding solution a net exchange between resin and solution takes place, which in case of univalent anions or cations can be described by the following two equations:

$$\overline{AR^+A_1^-} + A_2^- \rightleftharpoons \overline{AR^+A_2^-} + A_1^-,\qquad(2.34)$$

and exchange of univalent cations on a CR can be written:

$$\overline{CR^-M_1^+} + M_2^+ \rightleftharpoons \overline{CR^-M_2^+} + M_1^+,\qquad(2.35)$$

where $\overline{AR^+}$ and $\overline{CR^-}$ refer to fixed ions in an anion or a cation-exchange resin, respectively and A^- and M^+ refer to anions and cations; the subscripts 1 and 2 indicate two different species of ions and the upper bar indicates the phase of the ion-exchange resin.

In ion-exchange equilibrium both kinds of ions are present in the resin as well as in the solution, and the ratio between their concentrations is described by an ion-exchange equilibrium constant:

$$K_{A_1/A_2} = \frac{a_{A_1} \cdot \overline{a}_{A_2}}{\overline{a}_{A_1} \cdot a_{A_2}} \text{ and } K_{M_1/M_2} = \frac{a_{M_1} \cdot \overline{a}_{M_2}}{\overline{a}_{M_1} \cdot a_{M_2}}.\qquad(2.36)$$

In Eq. (2.36) the equilibrium is defined through the activities of the components in the different phases. Since it is very difficult to determine activities inside the resin phase, ion-exchange equilibrium constants are more often defined through the molar concentrations or through the equivalent fractions of ions. Such equilibrium constants are usually concentration dependent and an ion-exchange isotherm, covering the required range of concentrations, can be used to describe the ion-exchange equilibrium, as illustrated in the **Fig. 2.7**, for the ion-exchange between H^+ and Zn^{2+} ions.

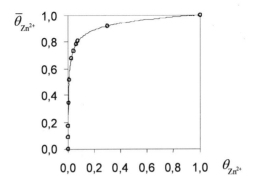

Fig. 2.7. Isotherm of Zn^{2+} absorption from 0,05 N $ZnSO_4/H_2SO_4$ mixture in the cation-exchange resin KU-2-8 (H^+), which is sulfonated polystyrene-*co*-DVB (8%), with the ion-exchange equilibrium constant calculated from the equivalent fraction of ions (θ)

$$^\theta K_{Zn^{2+}/H^+} = \frac{\overline{\theta}_{Zn^{2+}} \theta_{H^+}^2}{\overline{\theta}_{H^+}^2 \theta_{Zn^{2+}}} = 234 \pm 5$$

The preference of an ion-exchange resin for one kind of ions compared to others is called selectivity. Thus, the ion-exchange absorption isotherm shown in the **Fig. 2.1** indicates a higher selectivity of the resin to Zn^{2+}-ions compared to H^+-ions. For a strongly dissociated ion-exchange resin the selectivity depends mostly on electrostatic and steric interaction between counter-ions and fixed ions in the cross-linked structure.

The electrostatic interaction correlates with the surface charge of a hydrated counter-ion, which is a function of the charge and the diameter of the ion. Due to the electrostatic attraction with fixed ions, the counter-ions with higher charge number are more preferentially absorbed by the resin. This effect is stronger in diluted solutions and is referred to as electroselectivity of ion-exchange resins. Due to limits of space inside the swollen gel structure of an ion-exchange resin the counter-ions with smaller sizes are more preferably absorbed, compared to larger ions of the same charge. Also physicochemical interaction, like chelating or covalent bonding, plays an important role in selectivity.

Selectivity sequences of ions for different ion-exchange resins can be found in the literature [31,32,33] and in the specifications of some resin manufacturers. For standard gel-type, strongly acidic cation-exchange resins with 8% DVB, a typical selectivity row, indicating the increasing preference to absorb a cation, is:

$$Li^+ < H^+ < Na^+ < NH_4^+ < Mg^{2+} < Zn^{2+} < Co^{2+} < Cu^{2+} < Cd^{2+} < Ni^{2+} < Ca^{2+} < Sr^{2+} < Pb^{2+} < Ba^{2+}$$

This list shows that an ion-exchange of most salt cations is most efficient with a cation-exchange resin in H^+-ion form. Similarly, an exchange of most salt anions from the solution is most efficient with an anion-exchange resin in OH^--ion form.

The removal of acids with very low dissociation constants, like silicic and boric acid is due to the reaction of OH^--ions with the non-dissociated acid molecule. In general a weakly dissociated acid HA_w reacts with an anion-exchange resin according to:

$$\overline{AR^+OH^-} + HA_w \rightleftharpoons \overline{AR^+A_w^-} + H_2O. \qquad (2.37)$$

Weakly dissociated acids are often found in natural waters and their removal is an important issue in water desalination. Traces of carbonic acid are frequent anions in natural waters. The silicic and boric acids are present usually in lower amounts, but their removal is difficult due to the extremely low values of the dissociation constants. Weakly dissociated bases, like different alkylamines, are not very typical contaminants of natural waters and are more relevant to waste waters.

The rate of ion-exchange can be limited by diffusion of ions inside the ion-exchange resin beads or by diffusion of ions through the boundary layer of the solution at the interface with the resin. For many mineral ions in diluted solution and for standard ion-exchange resins the diffusion in the solution boundary layer is considered to be a rate limiting step. In both cases the smaller particle size results in a higher rate of ion-exchange. Ion-exchange resins with uniform particle size show a better performance in ion-exchange, compared to polydisperse resins with the same average particle size, the main reasons being the absence of big beads with slower kinetics, and the more evenly distributed flow through the bed.

2.2.4   Electrochemical properties of ion-exchange resins

Ions, not fixed in the gel structure of ion-exchange resins are mobile and can migrate once an electric potential gradient is applied. Thus, the resins are electrically conductive and are referred to as solid polyelectrolytes.

The conductivity of resins has different distinctive features compared to the conductivity of electrolyte solutions. The main mobile ions in a resin are the counter-ions which are present in a swollen resin in relatively high concentration (about $1 - 2{,}5$ eq/L for standard resins). The influence of co-ions on conductivity is often negligible due to their low concentration in the resin compared to counter-ions. The conductivity of co-ions becomes important however if the resin is in contact with highly concentrated solutions, or in resins with low cross-linking, low ion-exchange capacity (*IEC*) or weakly dissociated fixed groups.

Migration of ions always results in a certain volumetric flow, related to the volume of the ions and their hydrated shells. In electrolyte solutions the volumetric flow due to an externally imposed electric current takes place in opposite directions for positive and negative ions and if the volumes are not equal, the difference is easily compensated by the convection of free water. Since an electrical current in an ion-exchange resin is transported primarily by counter-ions, the water in the hydrated shells as well as some surrounding water is transported with the counter-ions, resulting in a unidirectional electro-osmotic flux.

Due to the strong steric restrictions of the gel matrix of a resin and due to electrostatic interactions between counter-ions and fixed ions, the mobility of an ion inside a resin is several times lower than in water. The mobility ratio $u_i/\bar{u}_i$ is higher for the counter-ions with higher charge numbers because they are stronger attracted by fixed ions (see **Table 2.3**). In special cases a physicochemical interaction between counter-ions and fixed ions, like formation of a complex bond or a suppression of fixed group dissociation can result in more than one order of magnitude decrease of the ion mobility and conductivity of a resin.

For ion-exchange resins with the same number of fixed groups per unit of the polymer chain an increase of the polymer cross-linking will decrease the water content in the swollen state and will bring the ions closer together. This leads to the simultaneous increase of the concentration of fixed ions, the ion-exchange capacity (*IEC*), and a decrease of the ion mobility ($\bar{u}_{cou}$). Both, IEC and $\bar{u}_{cou}$, affect the conductivity of the ion-exchange resin according to Eq. (2.26). Due to the different effects of cross-linking on the IEC and the $\bar{u}_{cou}$, the dependence of conductivity κ on the degree of cross-linking passes through a maximum.

In contact with water and dilute solutions the concentration of ions in the resin varies only slightly with the solution concentration. So, the conductivity of an electrolyte solution starting close to zero increases almost linearly with concentration, while the conductivity of the resin bed in equilibrium with this solution is initially much higher but has a lower increase. The intersection point of these two lines is called iso-conductivity point, which can be considered a characteristic of the specific ion-exchange material. An illustration for a strongly acidic CR with 8% DVB is given in **Fig. 2.8**, where the conductivity for an ion-exchange resin bed in contact with an electrolyte is compared for different concentrations of the electrolyte.

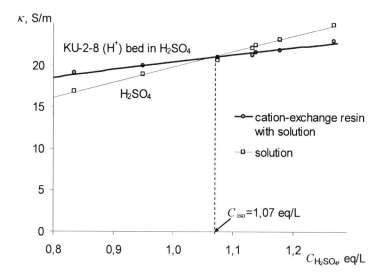

Fig. 2.8. Conductivity of H_2SO_4 solution and a bed of strong acidic cation-exchange resin KU-2-8 (H^+) with 8% DVB in equilibrium with H_2SO_4 solution at different concentrations, where iso-conductivity at the intersection point is indicated

In another method for the measurements of the electrical conductivity of a swollen ion-exchange resin bed, the solution is replaced by a non-conductive medium, e.g. by air. For that purpose a so-called centrifuge conductivity cell can be used, which should be calibrated for the specified ion-exchange resin using a resin sample with known conductivity, e.g. measured at the iso-conductivity point.

The conductivity of a strong acidic cation-exchange resin was measured with a centrifuge conductivity cell as described in [33] and the values for different mono-ionic forms are summarized in the **Table 2.3** and for bi-ionic Zn^{2+}/H^+ form in the **Fig. 2.9**. A sulfate solution with a concentration of 0,05 N was equilibrated with a resin before centrifugation and the conductivity was measured at 20°C.

Table 2.3

Conductivity, capacity and ion mobility of cation-exchange resin KU-2-8

in different ionic forms[*]

Counter-ion	$\bar{\kappa}_i$, S/m	$IEC,$ eq/L$_{(swollen)}$	$\bar{u}_i \cdot 10^9$, $\left(\dfrac{m^2}{V \cdot s}\right)$	$u_i^0 \cdot 10^9$, $\left(\dfrac{m^2}{V \cdot s}\right)$
H^+	11,1	2,87	40	363
Na^+	2,35	2,98	7,8	51,8
½ Zn^{2+}	0,67	3,02	2,3	55,4

[*]Ion mobility in water in the limit of zero concentration (u_i^0) is given for comparison

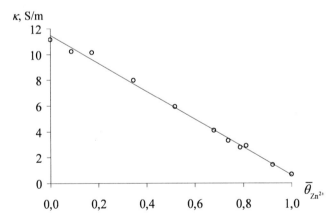

Fig. 2.9. Conductivity of cation-exchange resin KU-2-8 in bi-ionic Zn^{2+}/H^+ form as function of Zn^{2+} equivalent fraction

Alternatively, a resin equilibrated with water of low conductivity can be used for conductivity measurement. S. Thate [193] used water with $\kappa < 1$ µS/cm flowing through a bed for the measurements of the resin conductivity, assuming that the influence of the water conductivity on the measurements of the resin conductivity is negligible at this conditions. He showed that the area of the contact points between the resin beads have a strong influence on the bed conductivity. In this case small variations of the packing density or the pressure applied upon the bed can significantly affect the values of the bed conductivity.

In several publications describing the investigation of transport properties of ion-exchange materials [34-37] deviations from the Nernst-Einstein Eq. (2.16) are indicated. The mobility of counter-ions calculated from conductivity measurements (Eq. (2.26)) gives usually higher values than the mobility calculated from diffusion coefficients (Eq. (2.16)) which have been measured by alternative methods. As indicated in the Chapter 2.1.3.1 the Nernst-Einstein relation is derived in the limit of zero concentration, where ions together with their solvation shell move independently and are not influenced by charged species in their neighborhood. These conditions are obviously not present in ion-exchange resins.

2.2.5 Commercial ion-exchange resins

Commercial polystyrene based strong acid cation-exchange resins are applied in water softening as well as in water deionization, where they are usually combined with strong base anion-exchange resins. Most common are resins with about 8 wt.% of DVB as the cross-linking agent and sulfonic cation-exchange groups $-SO_3^-$.

Strong base anion-exchange resins are classified in resins of Type-1 and Type-2. The main fixed group for strong base anion-exchangers is the quaternary ammonium group $-N(CH_3)_3^+$, and respective resins are referred to as Type-1. The chemical structure of this resin type is shown in **Fig. 2.10***a*. Due to the high dissociation degree and ion-exchange capacity, these resins are very suitable for the general ion-exchange processes. In OH⁻ion form these resins have also very good affinity to weak acids like carbonic or silicic acid.

The anion-exchange resins of Type-2 contain quaternary alkanolamine groups as shown in **Fig. 2.10***b*. They are usually synthesized by the reaction of polystyrene-*co*-divinylbenzene with dimethylethanolamine or some polyol-amines. Type-2 resins are not as strongly basic as Type-1 resins, but they show sufficient removal of weak acids and are easy to regenerate.

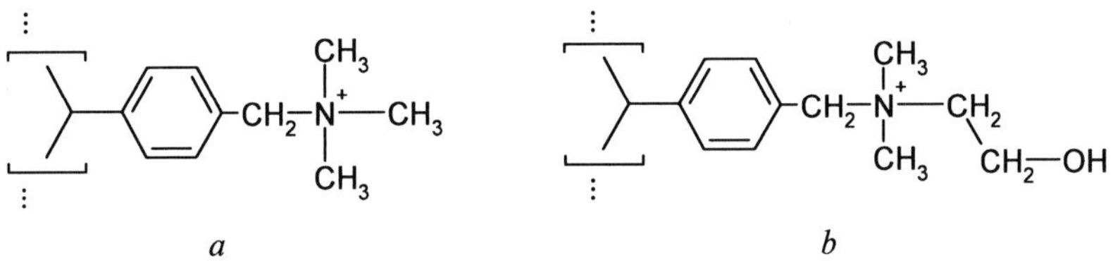

Fig. 2.10. Typical fixed ions of Type-1 and Type-2 anion-exchange resins bond to the benzyl group of matrix polymer: *a* – Trimethyl-ammonium group (quaternary ammonium group) of a Type-1; *b* – (Hydroxy-ethyl)-Dimethyl-ammonium group of a Type-2

In the deionization of water with ion-exchange columns silicium and boron from the respective weak acids appear first in the outlet of the column when the capacity of a column is exhausted and thus, can be used as indicators of the capacity of an ion-exchange column. Due to the very low dissociation constant (pK_1= 9-10) the removal of silicic and boric acid is only possible by the OH⁻ form of an anion-exchange resin or by chelating resins which selectively bind these elements.

In special cases the use of selective ion-exchange resins, such as an anion-exchanger loaded with metal hydroxide, can improve the silica removal. The removal of silica by ion-exchange can also be hindered if silica is present in a polymeric form.

The anion-exchange resin containing polyvalent alcohol groups as shown in **Fig. 2.11a** can selectively bind boron. These resins are weakly basic and the removal of boric acid takes place through chelate bonding of boron with the oxygen atoms of polyvalent alcohol groups as illustrated in **Fig. 2.11b**.

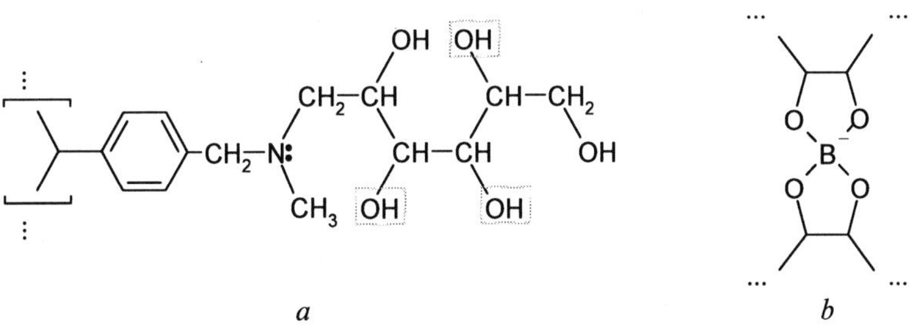

Fig. 2.11. Chelating anion-exchange resin for boron removal: *a* –N-methylglutamine group bond to benzyl group of a matrix polymer; *b* – complex of B(OH)₃ formed with two functional groups of polyol [38]

Today, different companies produce such boron selective resins. The use of selective ion-exchange resins can be more efficient in combination with standard resins.

In [39] it is proposed to place boron selective resins preferably at the end of all other purification steps.

Also different other kinds of resins, e.g. weak acid or weak base resins as well as resins selective to certain ions such as nitrates, heavy metals, etc., are now commercially available and used for specific applications.

2.2.6 Packed bed ion-exchanger

Ion-exchange processes are widely used in the water treatment. The most important applications are water softening, dealkalization, deionization or polishing. The bed of a cation- or an anion-exchanger or a mixture of both is packed usually in cylindrical vessels. While the water passes through the bed, ion-exchange between water and resin takes place.

In water softening, a cation-exchange resin is originally in form of a univalent salt ion (usually Na^+-ion) and during the treatment these ions will be replaced by the hardness cations from water. For deionization, a cation-exchanger should be in H^+-ion form and the anion exchanger in the OH^--ion form. During water treatment these ions will be replaced by mineral ions from water and the released OH^-- and H^+-ions combine to water.

At certain extend of saturation of the ion-exchange resin a breakthrough of the ions contained in the water occurs, and the resin must be regenerated or replaced. During water softening, a cation-exchanger is usually regenerated with saturated sodium chloride solution, while in deionization solutions with 2 - 8% acid or base are used. For mixed-bed columns the separation of cation- and anion-exchanger beads is required before regeneration.

The main disadvantages of conventional packed bed ion-exchange processes are their need of chemicals for the regeneration, their effluent of concentrated waste water, and the fact that packed bed ion-exchange is not a continuous process.

In Chapter 5 a novel continuous desalination processes will be discussed where beds of ion-exchange resins are placed between ion-exchange membranes and where the regeneration takes place via electrochemically generated H^+- and OH^--ions. For these processes the ionic conductivity of the ion-exchange bed is important. It will therefore be briefly discussed in the following.

2.2.7   **Ionic conductivity of ion-exchange resin bed**

If resin beads are present as a packed bed with an electrolyte solution in the void volume, the electric conductance of such beds depends on both, the resin and the solution [40]. As shown in **Table 2.3** and in the literature [31-33,41], the values of conductivity for strongly dissociated gel-type ion-exchange resins are mostly in the range 10^4-$10^5\,\mu$S/cm, with a strong dependency upon the counter-ion present. The conductivity of ultra-pure water is 0,055 µS/cm and that for typical RO-permeate produced from tap-water ranges within 10^1-$10^2\,\mu$S/cm. Thus the conductivity of the resin phase is 2 to 5 orders of magnitude higher than that of the surrounding water. An electric current will therefore be conducted almost completely through the ion-exchange resin, if parallel paths through the resin and the solution exist.

In a packed bed the direct current path through the beads is repeatedly interrupted by the contact points between adjacent beads. This makes the conductance of the packing sensitive to factors compressing the bed and influencing the contact area between beads. A packed ion-exchange bed should therefore be flown through from top to bottom to provide a sufficient pressure and prevent fluidization of the packing. In addition, if the ion-exchange bed is partly saturated and partly regenerated, its conductivity will vary with the local degree of regeneration.

2.3 Ion-exchange membranes

An ion-exchange polymer produced in form of a thin sheet or film without macroscopic pores is called an ion-exchange membrane (IEM). The main properties of an ion-exchange membrane are based upon the easy transport of counter-ions and the Donnan exclusion of co-ions. This allows counter-ions to pass easily through the membrane while co-ions are blocked as long as the ion concentration of the outside solution is sufficiently low. Ion-exchange membranes are therefore the key components in electromembrane desalination processes. A typical example is electrodialysis as illustrated in **Fig. 1.1**. Since the transport properties of ion-exchange membranes depend upon the nature of the ion-exchange material and the procedure of membrane manufacturing, production and morphology of the different types of ion-exchange membranes will be briefly reviewed.

2.3.1 Production of ion-exchange membranes

Different methods of production result in membranes with different structure. The main techniques applied in production of commercial ion-exchange membranes are:

1) preparation of a film from a polymer solution with the polymer having a positively or negatively charged moiety;

2) introduction of positively or negatively charged moieties into a polymer film by grafting, and chemical treatment;

3) formation of interpenetrating networks of an ion-exchange polymer with a neutral polymer; and

4) mixing of fine ion-exchange particles with a binding polymer.

The first two membrane preparation techniques result in a homogeneous distribution of the charged groups. The third and the fourth preparation procedure leads to heterogeneous membrane structure where the charged groups are accumulated in microscopic or macroscopic domains.

2.3.1.1 Homogeneous membranes

The preparation of a homogeneous ion-exchange membrane from an ion-exchange polymer solution includes usually the following steps:

1) casting the solution as a film on a smooth flat plate;

2) evaporation of solvent;

3) immersing of the obtained polymer film in water and detaching it from the plate; and

4) chemical or/and thermal treatment of the membrane (optional).

Often a cross-linking agent is added to the polymer solution and a cross-linking takes place during the solvent evaporation, or post-treatment.

Another method for producing homogeneous ion-exchange membranes is the grafting of a polymer film (PE, PVC, etc.) with suitable mechanical properties by chemical processes or by irradiation. Typically irradiation is used with styrene polymer films. The treatment is followed by a chemical functionalization of the benzene rings and the introduction of fixed ion-exchange groups [42,43]. Other polymer backbones, monomers, and functional groups can be used, depending on the desired property of the final membrane.

Synthesis of a homogeneous ion-exchange membrane is carried out in several steps and requires handling with toxic chemicals and expensive components. The production costs of homogeneous ion-exchange membranes are generally higher compared with heterogeneous membranes. Therefore they are used mostly in applications where a very high chemical stability or electrical conductivity is required such as in membrane electrolysis or in fuel cells. In electrodialysis homogeneous membranes are applied only in some special cases.

2.3.1.2 Microheterogeneous membranes

Different types of microheterogeneous ion-exchange membranes, e.g. interpenetrating network polymers, or block- and *co*-polymers are commercially available. Many of them consist of a mixture of an ion-exchange polymer with a carrier polymer and a plasticizer. A so-called "paste method" has originally been used for the production of such mixtures, casting them on an open mesh fabric in order to obtain a reinforced membrane [44-46]. Most of such commercial ion-exchange membranes for electrodialysis contain styrene-divinylbenzene *co*-polymer, chloromethylstyrene-divinylbenzene *co*-polymer or vinylpyridines-divivylbenzene *co*-polymer [47]. Also the membranes with an ion-exchange polymer structure based on acrylic polymer or other polymers are produced. The polymers of the mesh used for reinforcing ion-exchange membranes are mostly PVC, PE, PP and PA.

A possibility to manufacture polymer membranes as interpenetrating networks without formation of paste is described in the literature [48]. Here a film of a carrier polymer is swollen in a mixture of a liquid monomer and a cross-linking agent, like styrene and DVB. Then the polymerization results in the formation of an interpenetrating network of two polymers. The following functionalization with fixed groups produces a microheterogeneous membrane.

In block- and *co*-polymers the polymer chain contains parts with ion-exchange groups, which define the electrochemical properties and parts which define the mechanical properties of a membrane. By the adjustment of the type and the ratio of monomers during the synthesis, the membranes can be optimized in terms of their electrochemical properties and their mechanical and chemical stability.

In swollen microheterogeneous ion-exchange membranes the ion-exchange groups form hydrophilic clusters, which are somewhat larger than those of homogeneous membranes. Usually microheterogeneous ion-exchange membranes have sufficiently high conductivity and permselectivity as well as good mechanical and chemical stability.

2.3.1.3 Heterogeneous membranes

Heterogeneous ion-exchange membranes consist of two main components, i.e. a fine powdered ion-exchange resin and a binding polymer. If the fraction of ion-exchange powder is high enough to provide a continuous path of conducting particles from one membrane side to the other, the membrane is permeable for counter-ions and thus electro-conductive.

The technique of producing heterogeneous ion-exchange membranes from the mixture of a fine powdered ion-exchange resin with a binding polymer is known for more than 50 years [49-51].

The main steps in the production of heterogeneous ion-exchange membranes are:

1) grinding of ion-exchange resin;

2) mixing the powders of ion-exchange resin and binding polymer;

3) formation of a film from the mixture;

4) pressing of the film onto a reinforcing mesh (optional).

The step 1 is required to obtain a fine ion-exchange resin powder and is carried out by pulverizing the granulated ion-exchange resin in a mill, e.g. in a compressed air-jet mill. Some manufacturers of ion-exchange resins supply resins already in powdered form.

Step 2 is carried out in an extruder, in which the exactly portioned ion-exchange powder and the binding polymer are heated above a melting temperature of the binding polymer and then properly mixed. The ion-exchange powder used must be dry to prevent inhomogeneities and vapor formation inside the extruder.

Step 3 is carried out by calendering or pressing of the paste from the extruder into a film. The reinforcement of the membrane with fabrics and the final adjustment of the membrane thickness can be done during the step 4 by hot-pressing or calendering. Placing the fabric on both sides of the polymer film results in a symmetric reinforcement of the membrane which limits the deformation of the membrane due to swelling.

Inert thermo-plastic polymers are suited as bonding material in the production of heterogeneous ion-exchange membranes. They should exhibit the following properties:

- low glass-transition temperature ($< 20°C$);
- melting point higher than the maximal operating temperature of the membranes (80-$90°C$), but lower than the temperature at which ion-exchange polymers decay (90-$180°C$ depending on ion-exchanger);
- provide a good contact between the ion-exchanger particles;
- high chemical and mechanical stability; and
- low price.

In commercially available membranes polyethylene and polypropylene are usually used as a binder. Due to the availability and relatively low price of ion-exchange resins and binding polymers, the costs of heterogeneous membranes are significantly lower than those of homogeneous and microheterogeneous membranes. They are therefore very often applied in commercial electrodialysis systems.

2.3.2 Properties of heterogeneous ion-exchange membranes

The increasing membrane heterogeneity results in a decrease of the membranes price, but also in a degradation of properties relevant for practical applications, such as conductivity and permselectivity. Considering the microstructure, ion-exchange membranes can be divided into homogeneous membranes consisting of ion-exchange material only, and heterogeneous membranes, in which the ion-exchange material is embedded in an inert binding polymer [8]. Often, also an intermediate group of microheterogeneous membranes or other subgroups are distinguished for a more exact description. As mentioned before the structure of a membrane is directly related to the method of membrane manufacturing.

Even the homogeneous membranes prepared from a polymer solution can exhibit a certain inhomogeneity in the structure. For example in the homogeneous Nafion membranes the hydrophobic PTFE-based backbone forms a semicrystalline matrix and the fixed ions agglomerate into hydrophilic water-filled clusters of 3-5 nm in diameter depending on counter-ion [52,53].

In microheterogeneous membranes the formation of clusters is similar. Some parts of hydrophobic polymer and other parts of hydrophilic charged polymer form aggregates inside the interpenetrating network. Such clusters can be one order of magnitude larger than those in homogeneous membranes.

In heterogeneous membranes the dimensions of clusters, or gaps, are at least one order of magnitude larder than in microheterogeneous membranes and are defined by the size of ion-exchange particles as well as the properties and mass ratio of the components in the mixture with the binding polymer. The swelling of a heterogeneous membrane in water leads to a volume increase of ion-exchange resin particles as discussed in the Chapter 2.2.1, while a hydrophobic binder polymer does not change its volume. Hence, the swelling process can lead to a formation of gaps filled by water or solution, respectively as shown schematically in the **Fig. 2.12**. Such a swollen membrane includes the following three phases: 1) the swollen ion-exchange resin particles; 2) the hydrophobic binding polymer; and 3) micro-pores filled with water or electrolyte solution.

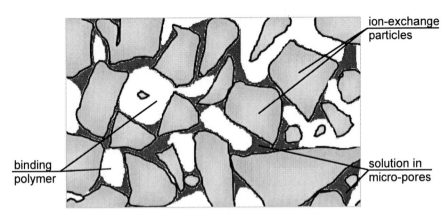

Fig. 2.12. Schematic drawing of the microstructure inside a swollen heterogeneous ion-exchange membrane

If the gaps are much larger than the Debye length, the ion concentrations inside the gaps should approach that of the outside solution, with equal numbers of cations and anions. If the potential gradient is applied across the membrane the current will be conducted through both the ion-exchange resin particles and the solution in the gaps. Since both cations and anions conduct the current in the solution, heterogeneous membranes are less permselective than a homogeneous or microheterogeneous membrane.

The conductivity of the membrane is determined by the distribution of the conductive particles inside a non-conductive phase and can be described by the percolation theory. The detailed description of basics, modifications and applications of this theory is given in the literature [54,55].

The basic approach of the percolation concept considers the number of particles contacting each other and building a continuous pathway in a cluster. Ions can only permeate a heterogeneous membrane if some clusters are large enough to cover the entire cross-section of the membrane.

Based on the percolation theory the cluster formation has been simulated. The software Perkol v.0.9. (author Lymar' S.S.) was used for cluster simulation and typical images of the largest cluster that cover the membrane cross-section are present in the **Fig. 2.13** for different fractions of particles.

44

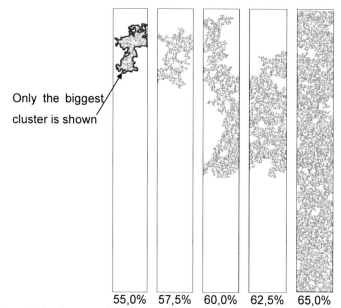

Only the biggest cluster is shown

55,0%   57,5%   60,0%   62,5%   65,0%

Fig. 2.13. Images of the biggest cluster across a membrane found for different volume fractions of ion-exchange particles (particle size 0,03 mm, membrane thickness 0,3 mm)

With a fraction of ion-exchange resin less than 55 vol.% only very few clusters cover the membrane thickness and the membrane will exhibit very low permeability. With increasing percentage of the resin in the mixture the number of clusters which occupy the whole cross-section of the membrane increases very fast, and with a resin percentage of more than 65 vol.% the ionic conductivity occurs across the entire cross-section of the membrane.

This example gives a rough estimation of the percentage of resin particles required for the production of membranes with a sufficient conductivity. For a heterogeneous membrane the fraction of resin particles should be high enough to exceed the percolation limit. However, a too high ratio of ion-exchange polymer to binder polymer reduces the mechanical properties of the membrane and its permselectivity due to excessive swelling.

Another point of concern in heterogeneous membranes is its surface structure which is often covered to a large extent by the hydrophobic binding polymer and only a small fraction of the membrane surface is permeable to ions. The extent to which the membrane surface is covered by the binding polymer is determined by the size and shape of the particles, nature of the binding polymer, the membrane preparation and the swelling of the ion-exchange particles inside the membrane structure. SEM pictures of the surface and the cross-section of a commercially available membrane used in the experimental studies of this work are shown in **Fig. 2.14**. The two micrographs show that the membrane is covered by a thin and rather dense skin with only very few macroscopic openings of a few micrometers. While the micrographs show definite structural differences between the bulk of the membrane and its surface they provide only little information as far as opening for the ion transport in the surface layer is concerned. Due to the sample preparation for SEM the membranes are completely dried while in any practical application the membranes are highly swollen expanding their volume by ca. 50%, that could increase the number of openings on the surface and its dimensions. Furthermore, due to the relatively low magnification only openings in the membrane surface with a diameter of more than a micrometer are visible. Thus, the SEM pictures of **Fig. 2.14** do not show the structure of the membrane under practical operating condition but they indicate differences in the distribution of the ion-exchange particles at the surface and in the bulk of the membrane.

Usually the choice of materials for the membrane production is based on material costs and their impact on the required properties. Thus, the use of ion-exchange resin with smaller particle size should lead to better mechanical stability and better permselectivity of the produced membrane, but its resistance might increase. The use of a binding polymer with a higher modulus of elasticity improves permselectivity and ion conductivity, but the film tends to be more brittle.

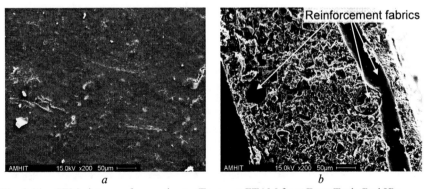

a *b*

Fig. 2.14. SEM pictures of a membrane (Fumasep FTAM from FumaTech GmbH): *a* – view of surface; *b* – cross-section prepared by breaking the membrane samples immersed in liquid nitrogen

2.3.3 Bipolar membranes

A special type of ion-exchange membrane is referred to as bipolar membrane. It consists of a cation-exchange and an anion-exchange layer, bonded together. A bipolar junction, also called transition region, is formed at the interface between the layers and is responsible for the water dissociation reaction. In some commercial bipolar membranes, the intermediate layer contains a catalyst such as weak bases or acids, or colloidal metal oxides for enhancing the water dissociation. The use of bipolar membranes and their function are discussed in more detail in Chapter 2.5.2.

2.3.4 Characterization of ion-exchange membranes

Standard characteristics of ion-exchange membranes include measurements of area specific resistance or conductivity, permselectivity, ion-exchange capacity, mechanical strength, water uptake and expansion by swelling. Different kinds of measurements are carried out to characterize the properties of ion-exchange membranes [131], as well as the properties of polymeric material and ionic components of the membrane [56,57].

Unfortunately, there is no international testing standard for determination of key properties of ion-exchange membranes, while different procedures and test conditions can lead to very different results. For example, the measurements of membrane conductivity with direct or alternating current result in different values, but even measurements with alternating current using impedance spectroscopy in different test cells cannot be directly compared [58]. For the experimental characterization of membranes in the present work conditions and methods close to those applied by membrane manufacturers were applied.

For determination of conductivity and area resistance the resistance of a membrane sample with known thickness and surface area need to be measured, which that can be done at conditions of direct or alternative current. For the measurements at direct current the membrane is fixed in the cell and rinsed with a solution, while current-voltage curve is recorded via e.g. Luggin-Haber capillaries, placed on both sides of the membrane. The resistance obtained from the linear increase of the current-voltage curve includes the resistance of the membrane and diffusion boundary layers of solution, which are depending on solution and flow conditions. Thus, the method includes some uncertainties in the determined membrane resistance and is difficult to reproduce in different kinds of test-cells.

For the measurements at alternative current the membrane under test is fixed in the cell between electrodes, being directly in contact with electrodes, e.g. mercury electrodes, or being separated from solid electrodes by a solution or other membranes [59]. In case when the membrane is not in contact with electrodes the membrane resistance is determined from the difference of the cell resistances with and without membrane. Since the solution between membrane and electrode can have resistance significantly higher than that of the membrane, it reduces the accuracy of the method and makes it inapplicable in dilute solutions. This limitation is eliminated when another membrane is used between the membrane under test and the electrodes. Such supporting membrane can be of the same as tested membrane, or another membrane type of the same polarity preferably with lower resistance. The method can also be modified to the measurements of the resistance of one, two, three, etc. stacked membranes under test and determination of its conductivity from the linear part in the dependence of the conductance upon the thickness of the membrane stack.

By the measurements at alternative current a constant frequency can be used, which is typically 1 kHz, or an impedance spectrum at different frequencies is measured and the active resistance is determined for the frequency corresponding to zero phase shift angle. Typically for the control of the membrane quality by manufacturing, the measurements at a constant frequency are sufficient, while for the comparison of the membranes of different types the measurements of impedance spectra are preferable.

For the determination of membrane conductivities in this work the cation- and anion-exchange membranes, as well as supporting membranes were transferred into Na^+-ion, or Cl^--ion form correspondingly and equilibrated with 0,5 M NaCl solution. Then a cation-exchange membrane was stacked between two Nafion 117 membranes, the stack was placed between electrodes, the impedance spectrum was measured and the active resistance was determined. The procedure was repeated for the stack of two supporting membranes alone and the resistance of the tested membrane was calculated as a difference between active resistances of both stacks. Membranes ADP were used as support by the measurements of anion-exchange membranes.

Permselectivity of ion-exchange membranes is usually characterized from the experiments of diffusion permeability of electrolytes or from the ion transport number measurements. By the measurements of diffusion permeability, the membrane in the cell separates two compartments with different concentration of the same electrolyte (or one compartment with water), while the changes of concentration in time are recorded and the diffusion permeability is determined from the linear part of that dependence.

Methods for the measurements of ion transport numbers through a membrane can be divided into two major categories [56]: a) measurements of ionic flux applying direct current (Hittorf's transport number); and b) measurements of electromotive force under a given concentration gradient over the membrane (potentiometric transport number). The determination of Hittorf's transport numbers is performed in a cell under constant electric current, when a solution is rinsing both sides of the membrane and the concentration changes are measured in time. Such measurements are sensitive to the conditions of the test affected by the diffusion boundary layers and other membranes or electrodes.

The measurement of the electromotive force is a static method where a potential difference appeared over the membrane placed between electrolyte solutions of different concentrations is measured. This potential difference is caused by the difference of Donnan potentials at both membrane sides and is maximal for the ideally permselective membrane for a given electrolyte concentrations and is lower for a membrane of limited permselectivity. Transport numbers obtained from the electromotive force measurements do not exclude the influence of water transport through the membrane and therefore are called as "apparent". These transport numbers are however not affected by concentration polarization and the measurements are fast, simple and easy to reproduce in different kind of cells.

For the measurements of the transport numbers in this work a membrane was equilibrated with 0,75 m KCl solution and placed in the cell separating two compartments rinsed by a KCl solution with molality 0,5 m and 1,0 m, thermostated at 25°C. The potential difference between reference electrodes immersed into the compartments of the cell was measured. Then the electrodes were exchanged between the compartments and the potential was measured again. The average of absolute values of both measurements ($\Delta\phi_{meas}$) was used to calculate the transport number using the formula:

$$t_{cou} = \frac{\Delta\phi_{meas}}{\Delta\phi_{max}}, \qquad (2.38)$$

where $\Delta\phi_{max}$ is the maximal potential difference for the used solutions:

$$\Delta\phi_{max} = \frac{RT}{|z_{cou}|F}\ln\frac{a_{\pm}^{m_2}}{a_{\pm}^{m_1}}, \qquad (2.39)$$

where $a_{\pm}$ is the average ion activity, T – temperature, z_{cou} – counter-ion charge number, R – universal gas constant, F – Faraday constant and superscripts m_1 and m_2 indicate solutions with lower and higher molality.

As an example the values used for the calculations obtained from a Nafion 117 membrane are listed below:

KCl molality		C, mol/L	$\gamma_{\pm}$	$a_{\pm}$	$\Delta\phi_{max}$, mV	$\Delta\phi_{meas}$, mV	t_{K^+}
m_1	0,5 mol/1000g H_2O	0,492	0,649	0,3245	15,96	15,44	0,967
m_2	1,0 mol/1000g H_2O	0,970	0,604	0,604			

Expansion by swelling and the water uptake can be determined from the measurements of linear dimensions and mass of the membrane in the swollen and the dry state. For the membranes tested in the present work two 10 cm long segments perpendicular to each other were marked on the dry piece of the membrane and the thickness was measured, while the location for the thickness measurement was marked. Then the membrane was immersed into water for swelling longer than one day and the length of the marked segments, the thickness in the specified location and the mass of the membrane were measured. The expansion by swelling or the mass uptake is calculated as ratio of a corresponding dimension or mass of the swollen to that of the dry membrane.

Mechanical strength of the ion-exchange membrane can be characterized using the standard method ISO 527-3, which is suitable for the measurements of the bursting strength of flexible films thinner than 1 mm using a test specimen of Type 4. In case of measurements performed on swollen membranes, the test procedure of the above mentioned standard need to be modified for preventing the test specimen from drying during the measurement. For this purpose the specimen of membrane can be enveloped by a wet porous paper, which has negligible bursting strength.

For the measurements of the total ion-exchange capacity of the membranes standard methods used for the ion-exchange resins can be applied.

2.4 Ionic transport in electromembrane processes

Ion-exchange membranes are usually applied in membrane processes like electrodialysis, introduced in **Fig. 1.1**. Here the membrane process consists of a sequence of compartments, confined by cation- and anion-exchange membranes and flown through by solutions of different ionic concentrations. The arrangement of the membranes and the electrical potential is such that cations and anions migrate perpendicular to the flow direction from the diluate compartments through the cation- and anion-exchange membranes into the adjacent concentrate compartments. As explained before, this ion migration is accompanied by some water transport. In the concentrate compartments the further migration of cations is stopped by an adjacent anion-exchange membrane and of anions by an adjacent cation-exchange membrane, leading to an accumulation of ions.

Migration and diffusion of ions and neutral molecules in electromembrane processes can be described by the general transport equation Eq. (2.13). If pressure diffusion and temperature gradients are neglected and convective mass flow is added, the general transport equation results in:

$$-J_i = D_i C_i \nabla(\ln \gamma_i) + D_i \nabla C_i + \frac{z_i}{|z_i|} u_i C_i \nabla \phi + C_i L_h \nabla P . \tag{2.40}$$

where J is the flux, D – diffusion coefficient, C – molar concentration, T – temperature, P – pressure, γ – activity coefficient, z – ion charge number, ϕ – electric potential, L_h – hydrodynamic permeability, R – universal gas constant.

The last term in Eq. (2.40) describing the pressure driven flux through a membrane material, as in the extended Nernst-Planck equation Eq. (2.18), assumes a linear dependency between the hydrodynamic flow and pressure gradient. Eq. (2.40) can be applied to the ion-exchange membranes and to the electrolyte solutions.

For a complete description of the flux of a component through a membrane separating two solutions the transport in the bulk solution, in the solution of the diffusion boundary layers at the membrane surface (Nernst's layer), in the ion-exchange membrane and the electrical double layer at the boundary between the outside solution and the ion-exchange membrane must be considered, as illustrated in **Fig. 2.15**. The transport in each of these elements will be briefly reviewed in the following.

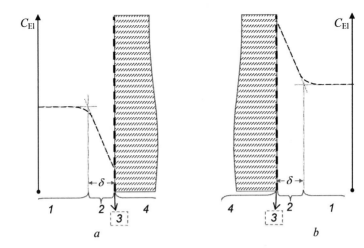

Fig. 2.15. Electrolyte concentration (C_{EI}) profile (dashed line) in the solution at: a – diluate side; b – concentrate side of an ion-exchange membrane. *1* – bulk solution; *2* – diffusion boundary layer; *3* – electric double layer; *4* – ion-exchange membrane

2.4.1   Transport in the bulk solution

The transport mechanisms in electrolyte solutions have already been discussed in Chapter 2.1.3. In electromembrane processes the bulk solution generally flows through compartments confined by ion-exchange membranes and filled with spacers such as a mesh screen or with a packed bed. As an example, two types of monofilament mesh screens are shown in **Fig. 2.16**.

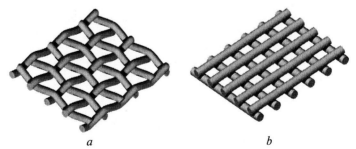

Fig. 2.16.   Drawing showing two types of open-mesh screens as an example of: a – woven screen; b – two-plane screen

The angle of the main flow direction to the filaments in a rectangle type spacer is generally 45°. The influence of the spacer design on pressure drop and turbulence promotion is discussed in different monographs [60-62]. In the channel spacers equalize concentrations perpendicular to the main flow direction. Thus, in the main flow direction only the last term of Eq. (2.18), responsible for convective flow, is of importance.

If the lateral mixing through spacers and flow turbulence is sufficiently strong, species concentration is uniform across the bulk of the solution and any diffusion transport resistance vanishes. This is however not the case for the laminar boundary layers close to the membranes surfaces.

2.4.2   Concentration polarization

The model of a laminar boundary layer of finite thickness is generally used to specify heat and mass transfer between the well mixed core (or bulk) of a flowing liquid and a confining wall. As shown schematically in **Fig. 2.15**, the concentrations C_{El} of an electrolyte are assumed to change linearly from the bulk value to a value at the membrane surface in the stagnant boundary layer of thickness δ.

Considering the diluate side of an ion-exchange membrane, the ion concentration will drop in the boundary as shown in **Fig. 2.15***a*. For counter-ions this drop is due to the lower transport number of the counter-ions in the liquid boundary layer compared to the ion-exchange membrane, while for co-ions the transport number in the ion-exchange membrane is lower than in the boundary layer. Electroneutrality enforces equivalent concentrations for co- and counter-ions in the boundary layer. The ionic concentration gradients in the boundary layers cause diffusive fluxes and the interplay of diffusion and electromigration results in the final form of the concentration changes. Similar arguments lead to the conclusion that in the boundary layer at the concentrate side of the membrane the ionic concentration falls from a maximum at the membrane surface down to the bulk value, as illustrated in **Fig. 2.15***b*.

Summarizing, the concentration changes in the boundary layers, referred to as *concentration polarization,* tend to increase the concentration differences of co- and counter-ions between both sides of an ion-exchange membrane, compared to the concentration differences of the adjacent bulk solutions. This means that the undesired diffusive flux of co-ions through the membrane is increased while the desired migration of counter-ions is decreased. A decrease of concentration polarization is therefore a goal of continuing importance in electro-membrane process development.

An approximate quantification of the concentration polarization can be obtained as follows: The ionic flux of each species must be the same through the boundary layer and through the adjacent membrane. If electromigration in the boundary layer and the ion-exchange membrane is described by transport numbers t_i and $\overline{t_i}$ (see Eq. (2.24)) and by diffusion (by third term in Eq. (2.15)), the following equation can be obtained for a binary 1:1 electrolyte [63]:

$$C_{El}^S = C_{El}^B \mp \frac{i \cdot \delta \cdot \left| t_i - \overline{t_i} \right|}{F \cdot D_{El}}, \qquad (2.41)$$

where C_{El}^S and C_{El}^B are the electrolyte concentrations at the membrane surface and in the bulk solution, i is the current density and D_{El} – the diffusion coefficient of the electrolyte in the solution. The minus-sign is valid for the diluate and the plus-sign for the concentrate side of the membrane.

Since at the diluate side $t_i < \overline{t_i}$ for counter-ions, the ionic concentrations drop with an increase of the current density until C_{El}^S reaches zero in Eq. (2.41). This current density is called the *limiting current density* (i_{lim}). In the above simplified picture, the limiting current density is the maximum current density to be applied in electro-membrane processes for ion removal. At higher current densities electrochemically enhanced water dissociation takes place at the membrane surface, which will be discussed in Chapter 2.4.5.

Concentration polarization can be influenced by the effective thickness of the diffusion boundary layer δ, which should be as thin as possible. For flow in flat, empty channels, correlations for the calculation of δ are available. But due to the complex flow pattern in a spacer-filled channel the thickness of the diffusion boundary layer can only be determined experimentally.

Limitations of the current density due to concentration polarization remain to be one of the central problems in electrodialysis. Numerous publications have been dedicated to this topic [64 - 72]. For a given limiting current density the total current can be increased by increasing the effective surface of the membrane, for example by using membranes with a corrugated surface geometry. In Chapter 3 a new approach based upon surface modification of ion-exchange membranes will be presented.

Another way to increase the effective surface area is to fill the channel adjacent to the membrane with ion-exchange media of the same polarity as the membrane, like ion-exchange screens, non-woven fibers or beads. In this case most ions are taken up from the solution by the ion-exchange media from where they migrate to the membrane. New processes based upon this principle will be presented and discussed in detail in Chapter 4.

2.4.3 Donnan potential and charge separation at the solution/membrane boundary

In most mathematical models describing the electric boundary layer between the outside solution and the ion-exchange membrane, ionic equilibrium between solution and ion-exchange material is assumed as explained in Chapter 2.2.2. This means that the electrical potential changes abrupt at the *solution/membrane* boundary by the amount of the Donnan potential (Eq. (2.33)). The abrupt junction is, of course, a simplification of reality. In a more refined picture [73], counter-ion concentration exceeds co-ion concentration in the solution close to the membrane surface, resulting in a space charge, i.e. electric double layer, which limits further charge separation. None of these models take possible heterogeneities in the surface structure into account, where clusters of fixed ions alternate with clusters of neutral polymeric backbone and microchannels filled with pore liquid. This results in a local variation of surface charges which becomes important in explaining phenomena occurring at overlimiting current density (see Section 2.4.5). In the present context however the assumption of a jump of the potential at the solution/membrane interface is sufficient.

2.4.4 Permeability and permselectivity of ion-exchange membranes

If the ion composition and the concentration of solutions on both membrane sides do not differ significantly, then swelling and counter-ion concentration across the ion-exchange membrane can be assumed as constant. In this case the concentration and potential gradients across the membrane are also constant and Eq. (2.18) can be applied to determine the ionic fluxes.

In electromembrane processes the ionic transport through an ion-exchange membrane takes place primary by migration. Therefore ionic permeability of ion-exchange membranes is usually characterized by the conductivity, which is mostly due to the migration of counter-ions as major mobile ionic component of the membrane (Eq. (2.26)).

Diffusion, pressure driven transport, as well as electromigration of co-ions from concentrate to diluate compartments is usually undesired. It can draw contaminations into the purified water stream and can be considered as process limitation in pure water production. Generally two types of species penetrating the membrane and contaminating the diluate can be differentiated: a) co-ions present in charged form; and b) neutral molecules such as a non-dissociated form of electrolytes. In the processes where a high concentration in the concentrate need to be achieved the water transport from the diluate into the concentrate is a limitation.

The penetration of neutral molecules through an ion-exchange membrane, such as the non-dissociated form of weak electrolytes, is not sensitive to the electric potential and not influenced by the Donnan exclusion. It is only caused by diffusion, which in turn depends upon the concentration gradients. The uptake and the diffusion coefficient of neutral molecules inside ion-exchange membranes is limited mostly by steric restrictions and the affinity of the membrane material.

It should be noted that the dissociation degree of weakly dissociated electrolytes is decreasing with concentration. Thus, weak electrolytes at the membrane surface facing the concentrate compartment are present more in non-dissociated form than at the surface facing the diluate compartment and are therefore able to penetrate the membrane and diffuse into the diluate.

A mathematical simulation of weak electrolytes diffusion through an ion-exchange membrane is complicated due to the difficulties to determine its dissociation constant inside the membrane. This is especially true for membranes containing some amount of H^+-ions or OH^--ions as counter-ions, which have strong influence on the dissociation of weak acids and bases.

It should be mentioned that by using heterogeneous membranes the undesired influence of back-diffusion from the concentrate into the diluate compartment can be counteracted to some extent by applying a pressure difference from the diluate into the concentrate compartment.

The pressure driven transport through the membrane is usually characterized by hydraulic permeability which values are extremely low for most of commercial ion-exchange membranes (10^{-10}-10^{-11} cm^{3}g^{-1}s^{-1} [74]), and is very dependent on membrane microstructure. Under the typical conditions of electromembrane processes no significant pressure difference is applied over membrane and the term of Eq. (2.17) and Eq. (2.11) describing the pressure driven diffusion $\overline{D}_i \dfrac{\overline{C}_i \overline{V}_{m\,i}}{RT} \nabla \overline{P}$ and the hydrodynamic flow $\overline{C}_i \cdot \overline{L}_h \cdot \nabla \overline{P}$ through a solid polymer electrolyte are usually minor comparing to other terms.

The experimental estimation of pressure driven water transport through the microheterogeneous type anion-exchange membrane AHA (Neosepta®) at 35°C and a pressure difference of 2 MPa resulted in the molar flux of $9 \cdot 10^{-3}$ mol(H_2O)m^{-2}s^{-1}. That corresponds to a hydraulic permeability of $1,5 \cdot 10^{-17}$ m^{2}s^{-1}Pa^{-1}, which is about three fold lower than that obtained in [75] for heterogeneous membrane at lower temperature (25°C). The higher hydraulic permeability of heterogeneous membranes is based on the flow through the gaps between ion-exchange particles and the binding polymer.

The total hydraulic permeability of a membrane can be considered as the sum of pressure diffusion and hydrodynamic flow, while the flux by the pressure diffusion can by estimated from:

$$-\overline{J}_{H_2O} = \frac{\overline{C}_{H_2O} \overline{V}_{m\,H_2O} \overline{D}_{H_2O}}{RT} \frac{\Delta P}{d}. \tag{2.42}$$

The rough estimation for the tested Neosepta®AHA is based on the following values: water concentration in the membrane $\overline{C}_{H_2O}$=16000 mol/m^3; partial molar volume of water $\overline{V}_{m\,H_2O}$=1,8·10^{-5} m^3/mol; diffusion coefficient of water in the membrane $\overline{D}_{H_2O} \approx 2 \cdot 10^{-10}$ m^2/s; pressure difference over the membrane $\Delta P = 2 \cdot 10^6$ Pa; and membrane thickness $d = 1,8 \cdot 10^{-4}$ m. Based on these estimations the hydraulic water permeability results in $4 \cdot 10^{-19}$ m^{2}s^{-1}Pa^{-1}, which corresponds to only few percent of the value obtained experimentally. Therefore it can be concluded, that the transport mechanism under the influence of a pressure gradient is of a mixed mode, in which the pressure diffusion takes a comparatively small part in the total water transport through the Neosepta®AHA membrane and the hydraulic water permeability of the membrane is mostly due to the convective transport through the inhomogeneity regions and the defects in the membrane structure.

Generally, the ratio between pressure diffusion and hydrodynamic pressure driven transport through an ion-exchange membrane depends strongly on the membrane structure. For homogeneous type membranes an impact of both transport mechanisms should be taken into account. For heterogeneous type membranes the role of pressure diffusion is negligibly low and only a hydrodynamic flow should be considered [76].

Permselectivity of ion-exchange membranes is usually characterized by the current transported by counter-ions versus the total current. The permselectivity is directly linked to the structure and the ion-exchange capacity of the membrane. For strongly dissociated ion-exchange groups the effect of Donnan exclusion decreases with decreasing ion-exchange capacity, resulting in increasing co-ions concentration inside the membrane (Eq. (2.33)). With an increase of membrane heterogeneity the fraction of micropores filled with solution becomes larger, which reduces the permselectivity for counter-ions.

The magnitude of the Donnan potential determines the exclusion of co-ions inside a membrane (see Chapter 2.2.2). Since the Donnan potential decreases with increasing ionic concentration outside the membrane (see Eq. ((2.33)), Donnan exclusion is reduced at the side of the concentrate compartment. As explained in Chapter 2.4.2, an electric current due to an applied electrical potential leads to concentration polarization at both sides of the membrane. This increases ionic concentrations at the concentrate side and reduces it at the diluate side of the membrane. As a result, the migrative driving force through the membrane is reduced for counter-ions and increased for co-ions. The concentration gradients between the concentrate and the diluate side of the membrane cause an additional diffusive flux, which also reduces the permselectivity.

Summarizing, transport through ion-exchange membranes is the result of a complex interplay of different processes parameters which is schematically illustrated in **Fig. 2.17**.

58

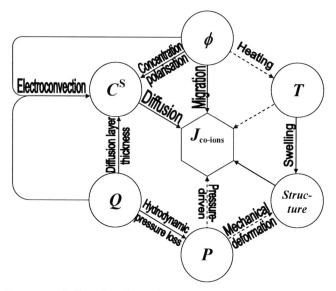

Fig. 2.17. Parameters influencing the co-ion transport in an ion-exchange membrane. -----▶ Dashed line indicates minor effects.

As indicated in the **Fig. 2.17** the concentration of co-ions in the solution at the membrane surface facing the concentrate compartment (C^S), the potential drop over the membrane, and the membrane structure are the main factors, directly influencing the co-ion transport. Other parameters like pressure and temperature have only a secondary effect.

The flow rate Q has no direct influence on the co-ions transport through ion-exchange membranes. It defines, however, the residence time of the solution in a compartment, i.e. the changes in concentration along the main flow direction. Additionally, the flow rate has a strong impact on the thickness of the diffusion boundary layers and on the pressure drop in the compartments. Taking this into account, the flow rate is influencing the cross-membrane co-ions transport through the other parameters.

The increase of temperature results in an increasing ion mobility and in a larger swelling of the membrane with a subsequent enlargement of the microchannels in the membrane structure. The swelling of the polymer network will reduce the charge density and hence the Donnan exclusion. In addition to the effects illustrated in **Fig. 2.17** the changes of the ion composition in the bulk solutions or inside the ion-exchange membrane will also change the membrane permselectivity. For example the uptake of

H$^+$- or OH$^-$-ions could result in a stronger membrane swelling, and hence a reduction of its permselectivity.

The permselectivity of an ion-exchange membrane is usually characterized by the membrane transport number, which is equivalent to the definition of a transport number for solutions (see Eq. (2.24)).

The flux of co-ions through a membrane for the example of cations penetrating an anion-exchange membrane can therefore be calculated by:

$$\left| J_+^{AM} \right| = \frac{\overline{t_+^{AM}} \cdot i}{z_+ F}.$$ (2.43)

2.4.5   Overlimiting current density regime and water dissociation

In addition to the homogeneous reaction of water autoprotolysis (Eq. (2.1)), the water dissociation can be catalyzed by an ion-exchange material. The catalytic region is located in the electric double layer at the *membrane/solution* interface, or the bipolar interface between cation- and anion-exchange materials, as discussed in Chapter 2.5.2 for the case of a BM [77]. The intensity of water dissociation increases with the catalytic activity of the reactive region, with a potential drop in the reactive region, and with temperature.

As mentioned in Chapter 2.4.2, water dissociation takes place at the diluate side of an ion-exchange membrane when the limiting current density is exceeded. For bipolar membranes, introduced in Chapter 2.3.3, this is the desired mode of operation. Here the "diluate side" is the microscopically thin transition region between the cation- and the anion-exchange layer. In bipolar membranes water dissociation is usually enhanced catalytically when the limiting current density is exceeded. The water dissociation in systems with ion-exchange membranes has been discussed in a large number of publications [71,78-80].

In the diluate compartment of an electro-membrane process the limiting current density depends on the ionic concentration of the bulk solution and also on the catalytic activity of the membrane surface which is usually higher at the surface of anion-exchange than at cation-exchange membranes. While the H$^+$-ions remain in the diluate and reduce the pH, the OH$^-$-ions migrate through the anion-exchange membrane towards the concentrate side, where they may cause membrane scaling through precipitation of hydroxides. Measurements of the current-voltage characteristic of electrodialysis cells accompanied by pH measurements carried out in [71] show, that water dissociation and pH changes occur, if a current higher than limiting current is applied.

Generally, electrodialysis is operated at current densities not exceeding 80% of the limiting current density to prevent pH changes in the diluate and scaling in the concentrate and to avoid the electric power consumption of water splitting. In some cases however electromembrane processes are intentionally operated at overlimiting current densities. This is based upon the experience that, contrary to the above simplified picture, ion removal can be increased in electrodialysis over the value obtained at limiting current density. The explanation is related to the surface inhomogeneities of ion-exchange membranes, mentioned already in Chapter 2.4.3. Theoretical considerations and detailed experiments show that these surface inhomogeneities lead to microconvection close to the membrane surface, which increases mass transfer over the limits predicted by a stagnant diffusion boundary layer [81].

In addition, OH$^-$- and H$^+$-ions produced by the water dissociation favorably influence the conditions close to the membrane surface. This can be explained by the example of NaCl removal from the diluate. Without electrochemically enhanced water dissociation and convective phenomena in the diffusion boundary layer, the limiting current density is determined by the combined diffusivities of Na$^+$- and Cl$^-$-ions in the diffusion boundary layer (see **Fig. 2.15**). If due to water dissociation, parts of the Na$^+$-ions are replaced by H$^+$-ions, the combined diffusivity increases and the limiting current density will be higher, but the overall energy efficiency is reduced.

2.5 Electromembrane desalination processes

The main electromembrane desalination processes of current practical relevance are:

1) electrodialysis (ED);

2) continuous electrodeionization (CEDI);

3) electrodialysis with bipolar membranes (EDBM);

4) electrochemical regeneration of ion-exchange resins (ECR).

These processes will be briefly described in the following.

2.5.1   **Electrodialysis and its application**

Electrodialysis (ED) is a membrane based separation process in which ions are transported through ion-exchange membranes under the influence of an applied electric field. A simple ED unit consists of two compartments, separated by cation-exchange membranes (CM) and anion-exchange membranes (AM) and placed between two electrodes as illustrated in **Fig. 2.18**. The diluate compartment is separated from the adjacent concentrate compartments by an AM towards the anode and by a CM towards the cathode. Applying voltage between the electrodes results in a migration of cations from the diluate compartment through the CM and of anions through the AM. In the concentrate compartment (CC) the further migration of the cations toward the cathode is restricted by the anion-exchange membrane and of anions by the cation-exchange membrane. The overall result is the removal and the concentration of ions in alternate cells, obtaining a diluate (ion depleted water) in the diluate compartment and a concentrate in the concentrate compartment. Since the unidirectional transport of ions through ion-exchange membranes is always accompanied by some convective transport of water, the diluate is losing some water which reduces the salt concentration in the concentrate.

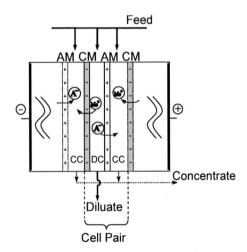

Fig. 2.18. Scheme of a simple electrodialysis unit. DC and CC are diluate and concentrate compartments; AM and CM are anion- and cation-exchange membranes respectively

Typically, electrodialysis is a continuous process, where the feed water is pumped through the ED-stack and is desalted during the passage through the diluate compartments. Usually, part of the feed water is used to rinse the concentrate compartments.

The most important application of ED nowadays is the desalination of brackish water to drinking water, which is realized at different scales from large scale desalination plants to mobile stations for construction sites or military camps. ED is also often used as a pre-treatment step in the production of boiler feed water, ultra-pure water, etc. Other areas where ED has found a practical application are:

- food and beverage industry - desalination of milk whey, fruit drinks, soy sauce, removal of tartrates from wine, etc.;
- drinking water conditioning - selective removal of nitrogen compounds (nitrate, nitrite, ammonium);
- agriculture - production of irrigation water;
- chemical industry - removal of ions from organic solvents, concentration of electrolytes, concentration control in automotive electrophoresis, etc.;
- potable salt production - concentration of sea water prior to evaporation;
- waste water treatment - recovery and reuse of water and valuable electrolytes.

2.5.1.1 Design and operation of electrodialysis

ED can be realized in plate-and-frame, as well as in spiral wound [82,83] type of apparatus. The plate-and-frame construction is the preferred system design and most widely applied. In large commercial ED stacks up to 500 membrane pairs are arranged between electrodes.

In steady-state the salt removal in electrodialysis can be described by a simple mass balance between the diluate stream and the concentrate stream of a cell pair (**Fig. 2.18**). The overall ion transport is proportional to the electrical current and given by:

$$\left(C_F - C_D\right)\cdot Q_D = \left(C_{CC^{OUT}} - C_{CC^{IN}}\right)\cdot Q_{CC^{OUT}} = \frac{\varepsilon \cdot I}{F}, \qquad (2.44)$$

where C is the equivalent concentration of the salt, Q is the flow rate, I is the current strength, ε is the current efficiency of the membrane pair, F is Faraday constant and the subscripts F, D, CC^{IN} and CC^{OUT} refer to feed, diluate, concentrate inlet and concentrate outlet respectively.

The voltage drop over a cell pair is the sum of the voltage drops over both membranes including their boundary layers and over the bulk of the diluate and the concentrate compartment.

Regarding the flow scheme of the ED cell, three main modes of operation can be distinguished: batch, continuous single-path and feed-and-bleed mode. The differences between the three modes are illustrated in the **Fig. 2.19**.

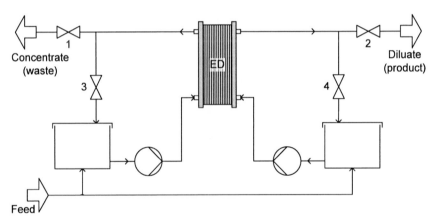

Fig. 2.19. Flow scheme of an ED stack operation, where in the batch mode the valves 1 and 2 are closed, while 3 and 4 are open; in the continuous single-path mode the valves 3 and 4 are closed, while 1 and 2 are open; and in the feed-and-bleed mode all valves are open and used to adjust the amount of recycle and the outlet flow (bleed)

The efficiency of water usage and the amount of waste water from ED is defined by the recovery rate ($\varLambda_{ED}$):

$$\varLambda_{ED} = \frac{Q_D}{Q_D + Q_C} \cdot 100 \ \% ,\qquad (2.45)$$

where Q is the flow rate and the subscripts D and C refer to diluate (product) and concentrate (waste) respectively.

Normally, the ED product is the diluate. In practical ED-desalination of fresh or brackish water is hampered by fouling and scaling of the membrane surfaces. These effects can be avoided to a large extent by electrodialysis reversal. In this process the polarity of the electric field is reversed in certain time intervals, simultaneously with a reversal of the diluate and concentrate flow streams, which removes fouling and scaling deposits from the membrane surface [84]. Flow reversal also reduces the need to add an acid, or anti-scalants into the feed water, allowing operation at high water recovery rate (up to 95%). The development of this technique has been a significant step in increasing the economics of the original ED and its competitiveness with reverse osmosis in brackish water desalination. A comprehensive description of main principals and technological details of ED can be found in different monographs [56,57,85].

The efficiency of ED-desalination is sufficient to provide drinking water. But the production of process water with very low ionic content is limited by concentration polarization in the membrane diffusion layers, by the high electrical resistance of water in the diluate compartment and by the limited permselectivity of ion-exchange membranes. In Chapter 3 an electrodialysis stack with profiled membranes is studied in detail.

2.5.1.2 Energy consumption in electrodialysis

The electrical energy consumption for the electrodialytical desalination process is proportional to the amount of salt that is removed from the feed water and transferred to the concentrate. In desalination of highly concentrated solutions or removal of specific ions the energy consumption is usually referred to the mass of removed salt or other target ions. For the production of fresh and pure water the energy consumption is usually referred to the volume of desalted water produced. Neglecting the water transport through the membrane, the energy consumption per volume E^V of produced water is:

$$E^V = \frac{IU}{Q_D},\qquad(2.46)$$

where I is the current, U the voltage and Q_D the diluate flow rate.

The current in the ED-stack depends on the number of cell pairs (n_{CP}) and the amount of ion equivalents removed:

$$I = \frac{F(C_F - C_D)Q_D}{\varepsilon\, n_{CP}},\qquad(2.47)$$

where ε is the current efficiency, C – the equivalent concentration of a specific ion or of salt, which is the sum of products of cation concentration and its charge number for all cations, and subscripts D and F refer to diluate and feed.

The total stack voltage is the sum of potential drops ($\Delta\phi$) at the electrodes and all cell pairs. Neglecting the role of electrode compartments, and assuming that cell pairs are identical, the stack voltage can be calculated by:

$$U = n_{CP}\Delta\phi_{CP}, \tag{2.48}$$

where $\Delta\phi_{CP}$ is the potential drop over one cell pair.

$\Delta\phi$ and ε can be estimated from characteristic parameters of the stack, or measured in laboratory scale cells. When the equivalent concentrations in feed (C_F) and diluate (C_D) are determined and $\Delta\phi$ and ε are estimated, the specific energy consumption related to the volume of the produced diluate can be found by combination of the equations (2.46), (2.47) and (2.48):

$$E^V = \frac{F\Delta\phi_{CP}(C_F - C_D)}{\varepsilon}. \tag{2.49}$$

It is evident from the Eq. (2.49) that at the same current efficiency and potential drop over the cell pair, E^V is directly proportional to the concentration difference between feed and diluate.

Regarding the energy consumption, electrodialysis requires less energy for brackish water desalination compared to most other currently used commercial desalination processes. ED systems are preferably utilized at *TDS* levels in feed water between 500 mg/L and 3000 mg/L [1,86], while reverse osmosis is preferred for desalination at higher and continuous electrodeionization (CEDI) at lower feed water concentrations.

66

2.5.2 Electrodialysis with bipolar membrane

Electrodialysis with bipolar membranes (EDBM) is mostly used for the production of acids and bases from a solution of the corresponding salts. EDBM is briefly discussed in the following since bipolar membranes will be applied in Chapter 5 for the continuous electrodeionization process with bipolar membranes.

A repeating unit of EDBM to be used for the production of an acid and a base is illustrated in **Fig. 2.20**. It consists of three individual cells containing the salt solution and the produced acid and base solutions, and three membranes, i.e. a cation-exchange, an anion-exchange and a bipolar membrane. A number of repeating cell units may be placed between two electrodes in a stack in such a way that the anion-exchange layer of the bipolar membrane is directed towards the anode and the cation-exchange layer towards the cathode. If an electric field between the electrodes is applied the OH⁻-ions generated in the bipolar membrane move towards the anode and form with the cations from the salt solution a base in the cell adjacent to the bipolar membrane. The H^+-ions generated in the bipolar membrane move towards the cathode and form an acid with the anions from the salt solution in the cell adjacent to the bipolar membrane. The two compartments containing the acid and the base are separated from the compartment containing the salt solution by an anion-exchange and a cation-exchange membrane, respectively.

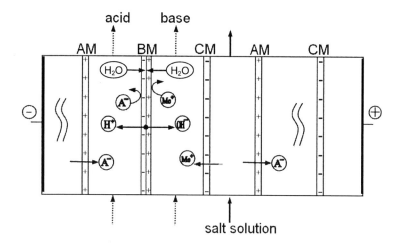

Fig. 2.20. Schematic diagram illustrating the acid and base production from the corresponding salt by electrodialysis with bipolar membranes. AM, CM and BM are anion-exchange, cation-exchange and bipolar membrane respectively

The overall process results in the production of an acid and a base from a respective salt solution. Water molecules dissociated in the bipolar membrane, are replaced by water from the surrounding solutions, diffusing through the two membrane layers to the transition region. Since the water dissociation rate is catalyzed by the ion exchange groups present, water diffusion through the two layers of the bipolar membrane may become the rate limiting step of the process. A limiting current density for bipolar membranes is therefore reached when the water concentration in the transition region of the bipolar membrane drops to zero. For usual commercial bipolar membranes this limiting current density is in the order of a few thousand A/m^2, which is significantly higher than typical current densities of conventional ED processes [87,88].

Different arrangements of bipolar and monopolar membranes in a stack are applied depending on the application. The comprehensive description of transport phenomena in EDBM, as well as the present and potential applications can be found in the literature [56,89-91].

2.5.3 Electrochemical ion-exchange

Some electromembrane processes use *IEC* of some materials to remove ions and the ion-exchange media will be regenerated electrochemically. Such processes can be performed in cycles of sorption/regeneration, or continuously with simultaneous processes of ion-exchange and regeneration.

One process performed in cycles is called "electrochemical ion-exchange" (EIX) [92]. In the desalination cycle of the process the migration of cations into a cation-exchange media and anions into an anion-exchange media is performed under an electric current, so that ions from solution migrate into the media and replace there the H^+- or OH^--ions, correspondingly. In the next step the polarity is reversed and ions in ion-exchange media are replaced by H^+- and OH^--ions and migrate into a concentrate stream, which is discharged to waste.

Usually the H^+- and OH^--ions are generated at the electrodes, which are covered by a layer of ion-exchange resin fixed with binder, or by an ion-exchange membrane. The use of bipolar membrane to generate H^+- and OH^--ions for the regeneration of membrane material is also applied [93,94].

EIX can be advantageous for the selective removal of specific ionic components like small quantities of some ions present in solution with high concentration of other

68

ions. EIX finds its application in the treatment of radioactive wastes as well as in some domestic water desalination apparatus.

Another form of a process performed in cycles is electrochemical regeneration of an ion-exchange resin, which was charged with ions during the previous ion-exchange stage [95]. In this case a resin is placed between IEMs of the same polarity or/and between electrodes and under the influence of applied electric current the ions in the resin migrate into the adjacent compartment and are discharged. These ions are replaced by the H^+- or/and OH^--ions produced electrochemically. Moreover, heavy metal ions removed from the CR can precipitate at the cathode as in electroplating, and thus can be reused in the production stage, for example, as a dissolved anode in an electroplating bath [96]. An example of electrochemical regeneration of ion-exchange resin charged with zinc ions with simultaneous precipitation of zinc at the cathode is illustrated in the Fig. 2.21.

Due to the relatively low current efficiency and less convenient discontinuous process the electrochemical regeneration in this form is currently not applied on a large scale, but only in some special small scale applications. Nevertheless, electrochemical regeneration can be realized simultaneously with ion-exchange removal, resulting in continuous operation and higher current efficiency, which makes it more attractive for industrial applications [97,98].

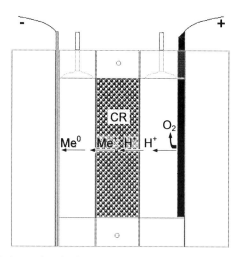

Fig. 2.21   Scheme of electrochemical regeneration of cation-exchange resin loaded by heavy metal ions with subsequent electrodeposition of the metal at the cathode

The process where ions removal on ion-exchange resin and electrochemical regeneration of ion-exchange resin occur simultaneously and results in removal of both, cations and anions, relates to continuous electrodeionization.

2.5.4 Continuous Electrodeionization

Continuous electrodeionization (CEDI) is an electromembrane process where diluate compartments are filled with ion-exchange material. In original form CEDI is the combination of ED with conventional ion-exchange technology and was applied to water deionization. The first CEDI modules were basically reproducing ED stacks (**Fig. 2.18**) with the diluate compartments filled with mixed-bed ion-exchange resins, as illustrated in the **Fig. 2.22**. The use of an ion-exchange resin, the conductivity of which is much higher than that of diluted solutions, allows the treatment of solutions of very low electrolyte concentration at sufficiently low operating voltage. Electrochemically regenerated ion-exchange resin is also applied for the removal of strongly and weakly dissociated electrolytes, resulting in completely deionized water as the product.

CEDI technology is successfully commercialized for the deionization on different scales and for different applications. No use of chemicals and continuous operation makes CEDI advantageous in competition with conventional ion-exchange.

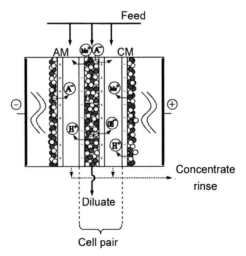

Fig. 2.22 The concept of a CEDI-stack with mixed-bed ion-exchange resins in the diluate compartment

A lot of modifications of original CEDI, optimization of process parameters, development of new principal concepts, etc., were done during last decades and resulted in an improve performance of the technology, in a broader application spectrum and in reduced investment costs of the CEDI modules and systems. Because of all this facts CEDI is currently one of the most rapidly developing electromembrane technologies in water deionization.

The research in this field is very application-oriented. It is carried out mostly by companies manufacturing CEDI modules, or CEDI including water purification systems. The results of such research are generally not published but remain as know-how. Therefore there is a lack of published fundamental research and systematic investigations of CEDI. More theoretical and experimental work is required to describe the influence of water dissociation, membrane properties, type of ion-exchange media, principals of module construction and secondary phenomena occurring in CEDI.

A more detailed discussion of CEDI is given in Chapter 4.

2.5.5   Other electromembrane processes

An electromembrane separation process, where electrode reactions are effectively used in the process scheme, is called electro-electrodialysis (EED). Water splitting, metal precipitation or ionic reduction-oxidation at the electrodes is mostly used as reactions involved in the process scheme. Such EED can be considered as a combination of membrane electrolysis and ED.

Depending on application a configuration of EED can contain only one membrane like in membrane electrolysis, but can also have two membranes of different polarity. The cells in EED consist usually of only 2 or 3 compartments and are therefore limited to small-scale operations. EED is a potential technology for a wide spectrum of applications from analytical chemistry [99,100] to hydrometallurgy [101]. Main industrial applications of EED aim at the recovery of different acids from wastes [102 - 107]. On a larger scale EED can often be replaced by ED with bipolar membranes.

Capacitive deionization [108] is another desalination method, which operation principal is close to electrochemical ion-exchange: it is a discontinuous process, where the same compartment is used in cycles for desalination and for concentration. The process can be realized with and without ion-exchange membranes used as charge barrier [109]. In both cases the ions from feed are removed by applying an electric potential difference, and captured at electrodes with very large specific surface areas, such as carbon aerogel. Due to the low energy consumption capacitive deionization seams to be very promising for the desalination market, but is currently limited due to the high material cost and requires further development.

Additionally, other very specific and still developing electromembrane processes like membrane reactors, electromatathesis, electro-ion substitution, etc., are known and their principles are described elsewhere [107].

Diffusion dialysis and Donnan dialysis are very often also referred to the electromembrane processes. Even if no external electrical field is applied to a membrane unit in these processes, the electric potential gradient rises from the concentration gradient of some ions and acts as driving force for electromigration of other ions. Diffusion dialysis is usually applied to recover acids or bases from their mixtures with salts. Donnan dialysis is not used in large industrial scale, but in some specific applications, like in analytical devices [56,110].

2.5.6 Combination of desalination processes

The required product quality in water desalination cannot be achieved by one process, and a combination of various technologies is required to meet a desired target. The choice of technologies for such combination is based on the feed and product water quality, such as defined in the Chapter 2.1.4, market segment, costs, etc.

The processes currently available for water desalination differ mainly in their principal, feed water requirements and desalination performance, as well as in their economics, i.e. operating and investment costs. Additionally, environmental aspects, footprints and other specific requirements play an important role for the selection of an optimal technology for a specified application. The abilities of existing water desalination technologies are intensively described in the literature [1,56,57,111].

Regarding technical feasibility a lot of differences can be found between the known processes. Distillation, RO and ED do not remove CO_2, silicates and some other substances completely from typical natural waters and are not able to achieve ultrapure water quality. On the other hand, ion-exchange or CEDI removes such contaminants to trace levels and are usually used in the last stages in ultrapure water production. RO is able to remove particular, microorganisms, viruses, and organic matters, but is more sensitive to fouling and scaling, etc. ED and ion-exchange do not remove particles, microorganisms and other biological contaminants efficiently.

For large scale applications, the energy costs are a large part of the overall operating costs, and they are therefore of special importance. A comparison of desalination costs done in [112] indicates that distillation is the most economic desalination process for feed water with very high salt concentrations, whereas, with decreasing salt concentration in feed water, other processes in the sequence RO, ED, ion-exchange, are more cost-efficient.

Combination of two or more processes in one system profits very often from the advantages of the individual processes and overcomes their disadvantages. In case of desalination with a large difference in the concentrations in the feed and product water the combination of technologies such as RO and ion-exchange is especially important and often used in practice.

The problem of large water desalination facilities to discharge a highly concentrated RO-retentate can be solved by the combination of RO with ED, whereby the ED stack comprises of ion-exchange membranes of conventional type and a type selective to univalent ions [113]. A combination of RO and ED in one system also helps improving the desalination performance and optimizing consumption [114].

In the production of process water with low conductivity, or UPW starting from tap or brackish water the use of ED or RO only is not sufficient. Currently, the combination of RO with CEDI has shown good results and is used more and more in industrial applications with high quality requirements for product water. In the combination of RO and CEDI, RO removes organic components and most of ions and CEDI removes the remaining ions and weak dissociated electrolytes. The CO_2 is not retained by RO, but can be efficiently removed by CEDI. In some cases however a degassing module can be used between the RO and CEDI processes [15,115] to remove CO_2 and reduce a load on downstream placed CEDI.

2.6 Conclusions and outlook

From the existing electromembrane processes, ED and CEDI are presently the most suitable for the production of low conductivity water.

The ability to produce low conductivity water by ED is limited due to the weak effect of the electric field on weakly dissociated electrolytes, the high electric resistance of the diluate solution, and the limited permselectivity of membranes. In spite of longer than 50 years presence on the desalination market, ED has a still a high potential for further development. The use of ion-exchange spacers, or membranes with special surface geometry, can improve the ability of ED to remove electrolytes and result in considerable energy savings and cost reduction [116].

The utilization of membranes with surface geometry different from a flat sheet can improve the performance of ED in the production of low conductivity water. The idea is to create membranes with a profile on its surface, allowing to reduce the electric resistance of the stack and to intensify the ions removal by increasing the membrane surface area and by providing good turbulence promotion and additional ion-exchange capacity for deionization. The selection of a suited surface profile, experimental tests of different methods to produce profiled membranes, as well as test results of ED with profiled membranes are present in the following Chapter 3.

CEDI, also called sometimes as "filled cell ED", is a more developed technique which has found a clear niche on the market. Due to the ion-exchange mechanism of ions removal, CEDI is able and also applied, to produce water with extremely low levels of residual ions, but requires higher investment cost than ED. Increasing demand of some industries for low-conductivity and stronger restrictions to the impurities content define the need of improvements in CEDI.

Detailed discussion of phenomena in CEDI and the description of the main existing CEDI concepts are present in the Chapter 4, while CEDI with bipolar membranes is selected from this comparison as a process, which is able to achieve the lowest level of ionic contaminations in the product water. Experimental investigations of different modifications of CEDI with bipolar membrane are presented in Chapter 5.

3 Electrodialysis with profiled membranes

As explained in the preceding section, the use of profiled ion-exchange membranes in electrodialysis stacks has the following advantages:

- the effective membrane surface area is increased;
- the concentration polarization effects are reduced by increased turbulence close to the membrane surface and closer location of membranes to each other;
- the stack construction is simplified since inert spacers between membranes are no longer needed.

3.1 Types of profiled membranes

Mechanical modifications of membrane surfaces can be differentiated depending on whether the whole membrane sheet is corrugated or only the surface of a flat membrane is structured. Regarding the scale of the surface modification it can be micro-structured, with surface structures significantly smaller than the membrane thickness, and macro-structured with structural dimensions comparable to the membrane thickness.

A membrane with a micro-structured surface can be considered as having a specific surface roughness [117,118]. A spacer is usually required for assembling such membranes into a stack, similarly to conventional flat membranes. A main reason for micro-structured surfaces is an increase of the surface inhomogeneity in order to create micro-convection which disturbs the diffusion layer at the membrane surface. Membranes with a micro-structured surface are claimed to offer a process intensification and better overall performance for the desalination of solutions with relatively high electrolyte concentrations, but could not solve the problem of high electrical resistance in diluate channels for the production of low conductivity water.

Membranes with macro-structured surfaces have structures substantially larger than 0,1 mm, and the membranes can either be corrugated or profiled. **Fig. 3.1** illustrates some possible examples of macro-structured membranes.

Corrugated membranes (**Fig. 3.1** *a*) are formed from flat membrane sheets by folding or pressing the membrane in a wavy shape. Different corrugated membranes, including ion-exchange membranes, were produced and tested by the group of K. Scott [119-125]. They tested corrugated membrane structures for filtration processes [119-121], where mass transfer coefficients in a channel with corrugated structures were determined [122,123]. The ion-exchange membrane Nafion$^{®}$117 was corrugated and tested in an electrolysis and electrodialysis cell without direct contact between adjacent membranes or electrodes [124,125]. In general, publications [119-125] demonstrate an increase of the mass transfer through a corrugated membrane compared to a flat one.

A corrugated ion-exchange membrane can easily be deformed by different factors such as different pressure in adjacent compartments or temperature. The installation of corrugated membranes in an ED stack therefore requires additional components, like special spacers to fix the membrane positions of adjacent membranes and to seal the membranes at the periphery. This complicates the application of corrugated membranes in ED stacks.

Profiled membranes, in contrary to corrugated, consist of a sheet with structural elements at the surface, which can have the shape of parallel linear notches as shown in **Fig. 3.1***b*, or of protrusions of different kind as shown in **Fig. 3.1***c*.

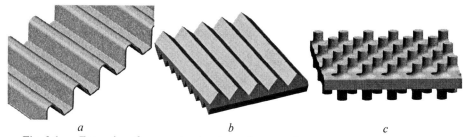

a   *b*   *c*

Fig. 3.1.   Examples of macro-structured membrane surfaces:
a – corrugated membrane [125]; *b* – profiled membrane with V-shaped notches; *c* - profiled membrane with protrusions [130]

An enlargement of the active membrane surface area and increased turbulent transport close to the surface are the main features of corrugated and profiled membranes, which both intensify the mass transfer between solution and the membrane surface.

Furthermore, the profiled membranes can be assembled into a stack without any spacers, directly point-contacting each other, while leaving a flow-through channel system between the contact points.

3.1.1　　　　Influence of the surface geometry on mass transfer

An optimal membrane profile should lead to an increase in the membrane surface area, to a good turbulence promotion and to a low hydraulic resistance in the flow channels formed. Some possible shapes of parallel notches on the surface of a profiled membrane are illustrated in **Fig. 3.2**.

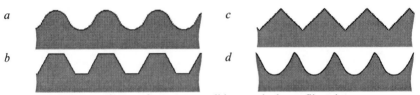

Fig. 3.2.　Cross-sections of some possible notched profile shapes: *a* – wave; *b* – trapeze; *c* – V-shape; *d* – saw-shape

It can be assumed that the direct stacking of profiled ion-exchange membranes of different polarity will have advantages for the desalination compared to a stack where flat membranes are separated by non-conductive spacers. The absence of non-conducting spacers increases the ion conductance across the channel and into the membrane by avoiding the so-called "shadow effect" of the spacer. Water dissociation will occur first at the contact points between adjacent membranes and the produced H^+- and OH^--ions will regenerate the respective ion-exchange membranes without changing the pH of the flowing solution.

The fact that the membranes in such a notched channel are close to each other (up to contact) should reduce the electric resistance of such a channel in diluted solution and provide more equal current density distribution along the channel. The diffusion boundary layers of both membranes will completely overlap at the perimeter of membrane contact, where the phases of water, as well as the cation- and anion-exchange membranes meet together. This creates advantageous conditions to intensify the salt removal.

78

Investigations of heat and mass transfer in systems with corrugated membranes have been presented by different research groups [61-62, 126-129]. These results help to select a suitable surface geometry for an optimization of mass transfer in ED stacks.

The profile geometry for the subsequent studies was selected based on the investigations of G. Gaiser [61] and F. Li [62]. A trapeze-shaped membrane surface profile (see **Fig. 3.2***b*) with notches oriented 90° to each other and 45° to the main flow direction was chosen for the experimental investigations. If subsequent membranes are placed on top of each other, crossing channels for the fluid flow are formed, as shown schematically in **Fig. 3.3**. In addition, a large number of contact points between the membranes of different polarity is formed, where water splitting can take place.

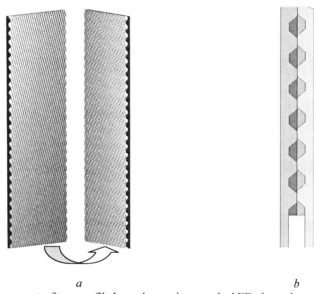

a   *b*

Fig. 3.3.   Arrangement of two profiled membranes into notched ED channel: *a* – orientation of notches to each other; *b* – cross-section of the two stacked membranes forming a flow channel

The ratio of the open membrane surface to the channel volume is an important parameter, characterizing the profile structure. Also the relative surface area of membrane contacts and perimeters of the contacts should be considered by the comparison of the profiles.

In the following, the dimensions of the notches have been chosen in such a way that their volume corresponds to the volume of the flow channels in conventional stacks with spacers (like **Fig. 2.16***b*) and no excessive pressure drop develops. Also, both sides of the membranes should be profiled so that adjacent membranes support each other.

3.2 Production of profiled membranes

3.2.1 Techniques to produce profiled membranes

Profiled ion-exchange membranes can be produced by different techniques, such as:

1) polymerization or polycondensation of liquid monomers in a mold provided with a suitable profile;

2) casting a solution of an ion-exchange polymer into a mold and evaporation of the solvent;

3) hot-pressing of a thermoplastic polymer containing suitable functional groups, or hot-pressing of a mixture of a thermoplastic polymer and an ion-exchange powder in a press-mold;

4) calendering (roll-pressing) of a thermoplastic ion-exchange film between corrugated rollers at elevated temperature.

The first of the above mentioned techniques requires a solvent-free polymerization of appropriate functional monomers. It can be carried out similarly to the paste method described in Chapter 2.3.1.2, but utilizing a profiled mold.

The second method can be used only if the ion-exchange polymer is soluble in a solvent or can be dispersed in a solution of a binding polymer. Contrary to the other techniques this procedure is suited to get the desired profile only on one membrane side at the mold surface, because the other side has to be uncovered in order to provide evaporation of the solvent.

The third and fourth methods can be applied to thermoplastic ion-exchange polymers or to mixtures of ion-exchange powder with a thermoplastic polymer. These techniques are easier to perform than the others and allow imprinting the desired profile on both sides of the membrane.

The surface of heterogeneous ion-exchange membranes contains significantly more neutral binding polymer than the bulk of the membrane. Therefore, the area available for the transport of ions is relatively small. This is well illustrated in the pictures from literature [70,130] and in the **Fig. 2.14** and **Fig. 3.10**. The phenomenon is related to the shape, size and percentage of ion-exchange particles in the membrane, but depends also on the pressing or calendering conditions. As shown in reference [130] at some pressing conditions profiling of heterogeneous ion-exchange membranes can result in stronger covered membrane surface, i.e. lower percentage of surface available for ion conduction.

Different materials and techniques have been tested in this work for the production of profiled membranes. Homogeneous membranes with a profile on one side were produced from solutions of ion-exchange polymers. For the production of membranes profiled on both sides, the pressing and calendering techniques were applied for mixtures of an ion-exchange resin powder and a thermoplastic polymer. Principally the paste method used for production of microheterogeneous ion-exchange membranes could be used alternatively for the production of profiled membranes, however, due to the high complexity of this method it was not utilized in the present work.

The procedures of membrane preparation and the properties of the membranes produced will be described in the following.

3.2.2   Profiled homogeneous membranes

Homogeneous membranes were prepared by using a polymer solution received from FumaTech GmbH. For the production of a cation-exchange membrane a dissolved sulfonated block-*co*-polymer was used; the resulting membrane had an ion exchange capacity of $IEC = 1,1$ meq/g.

For the production of an anion-exchange membrane a dissolved chloromethylated basic polymer was used, where strongly basic quaternary ammonium groups were introduced by treating with triethylammonium shortly before casting on the profiled mold. The IEC of the obtained membrane was 2,5 meq/g.

A mold with trapezoidal notches made of PTFE was used for casting the polymer solution and evaporating the solvent. The dimensions of the dry membranes produced in this mold are given in **Fig. 3.4***a* and *b*. In addition to the intended notches, volume shrinkage of the polymer solution during evaporation also leads to the formation of small corrugations on the upper, i.e. the gas contacting side of the membrane.

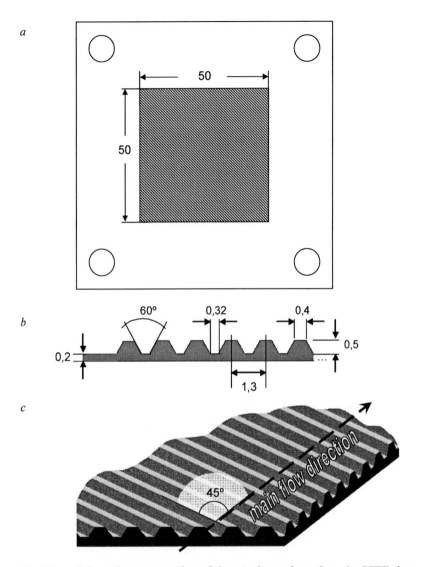

Fig. 3.4. Schematic representation of the membrane formed on the PTFE form with trapezoid notches at its surface with dimensions in mm: *a* - view of the membrane; *b* – cross-section of the membrane with the dimensions of the notches; *c* - position of notches to main flow in the channel

The active surface area of the notched membrane side is about 44% larger than that of a flat membrane with the same length and width. The dry ion-exchange membrane swells in water with corresponding dimensional changes as shown in **Fig. 3.5**. The

82

swelling of the membrane is somewhat non-uniform. It is slightly stronger in the area where more ion-exchange material is present, leading to a slight deformation of the notched shape.

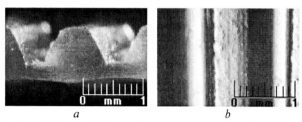

a   *b*

Fig. 3.5. Photos of the profiled membrane prepared from homogeneous polymer solution and swollen in water: *a* – cross-section of the notches; *b* – top view

The direction of notches is 45° to the main flow direction. Cation- and anion-exchange membranes are arranged on top of each other with notches crossing at an angle of 90° as shown schematically in **Fig. 3.3**.

As mentioned in Chapter 2.3, compared to heterogeneous membranes the homogeneous membranes have the advantage of higher ion-exchange capacity and ionic conductivity both inside the bulk and in particular at their surface. They have, however, also some disadvantages. The required polymer solutions are expensive, the preparation of the membrane by evaporation requires a relatively long time, and potentially hazardous chemicals must be handled. In addition, the evaporation of the solution is restricted to the production of membranes with a profile on one membrane side only. Therefore, only preliminary tests were carried out with homogeneous ion-exchange membranes.

3.2.3   Pressed heterogeneous membranes

For the preparation of heterogeneous membranes a commercially available ion-exchange resin was ground and the dry powder mixed with a polyolefin and pressed at elevated temperature into a heterogeneous ion-exchange membrane. The shape of the surface profile was defined by the press-mold. Several materials and pressing conditions have been tested in order to obtain a membrane with satisfactory properties for electrodialysis.

3.2.3.1 Grinding procedure

For the preparation of heterogeneous membrane it is preferable to use a fine powder with a particle size smaller than 0,1 mm (better < 20 μm) in order to prevent formation of cavity defects and to provide good mechanical strength of the membrane. Beads of conventional IERs were ground under different conditions and the particle size distribution of the powders obtained was measured.

Fig. 3.6 shows some measured particle sizes obtained with the optical analyzer Mastersizer 2000 with the dispersion module Hydro 2000G from Malvern Instruments. The measurements have been carried out with a dispersion of particles in water and show the size distribution of the swollen particles.

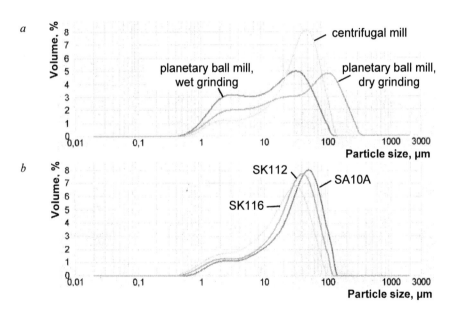

Fig. 3.6. Particle size distribution of powder swollen in water, produced by grinding of ion-exchange resins: *a* – comparison of different procedures, indicated in the graphs, by grinding the resin Dowex C400; *b* – grinding of swollen and frozen resins (Diaion) with different cross-linking degree in a centrifugal mill; DVB content in the Diaion resin SK116 is 16%; in SK112 – 12% and in SA10A – 10%

Fig. 3.6*a* is a comparison of powders produced from cation-exchange resin Dowex C400 by dry or wet grinding in the planetary ball mill PM 100 (Retsch GmbH), or by grinding of a resin, which is swollen in water and frozen in liquid nitrogen using the ultra-centrifugal mill ZM1 (Retsch GmbH) with a ring-sieve of 0,08 mm mesh size. Comparison shows that the centrifugal mill gives a narrower particle size distribution, while the planetary ball mill produces a big fraction of "dust" particles. The powders produced by wet-grinding in the planetary ball mill and in the centrifugal mill are acceptable for the membrane production. The grinding procedure in the centrifugal mill however, is less time consuming and was used in all further experiments.

Fig. 3.6*b* shows the particle size distribution of powders produced from Diaion ion-exchange resins with different degrees of cross-linking. Apparently, the mean particle size is slightly reduced with increasing cross-linking degree. This can be related to the higher rigidity of resin.

Further measurements of particle size distribution show that grinding of dry ion-exchange resins in the ultra-centrifugal mill results in slightly larger average particle sizes than by grinding of resins swollen in water and frozen in liquid nitrogen. It can be explained by the expansion of the resin volume by swelling and, possibly, by the higher rigidity of the water containing resins, frozen in liquid nitrogen.

The binding polymer was also ground in the ultra-centrifugal mill ZM1, using ring sieve 0,5 mm mesh size to provide an easier portioning into a mixture with a resin powder and to simplify the mixing of the components in the kneading device. The produced powder has an average particle size of ca. 0,2 mm.

Concluding, ion-exchange resins swollen in water as well as granulated polyolefins were frozen under liquid nitrogen, ground, dried and stored in a sealed container for the subsequent preparation of membranes.

3.2.3.2 Mixing of ion-exchange powder and binding polymer

The ion-exchange powder particles were equally distributed inside a molten binding polymer by kneading a batch of two-components. The masses of ion-exchange resin powder (m_{IER}) and of binding polymer (m_B) for the preparation of mixtures with desired volume fractions were calculated as following:

$$m_{IER} = \frac{\rho_{IER} \cdot \varphi_{IER}}{V_{mix}} \quad \text{and} \quad m_B = \frac{\rho_B \cdot (1 - \varphi_{IER})}{V_{mix}}, \tag{3.1}$$

where ρ is the density, φ – the volume fraction in the mixture, V_{mix} – the volume of the batch, subscripts IER and B refer to ion-exchange resin powder and binding polymer, respectively.

The densities of the binding polymers are taken from the data sheets of the suppliers. The density of the prepared dry powders of the ion-exchange resin (IER) was measured pycnometrically [131]. Isooctane was used as pycnometric liquid, because an IER powder does not swell in isooctane. The difference of masses of a given volume between isooctane, m_{IO} and of a dry resin powder, m_{IER} immersed into isooctane was used for the determination of resin density ρ_{IER}:

$$\rho_{IER} = \frac{m_{IER} \cdot \rho_{IO}}{m_{IER} + m_{P+IO} - m_{P+IO+IER}},\qquad(3.2)$$

where ρ_{IO} is the density of isooctane (692 kg/m^3 at 20°C), m_{IER} - the mass of IER powder, m_{P+IO} - the mass of the pycnometer with isooctane, and $m_{P+IO+IER}$ is the mass of the pycnometer with IER powder and isooctane.

For some samples the density was also measured with a gas-pycnometer. The results show a good correlation between both methods. The densities measured for some dry resins are: Dowex C500 ES UPN – 1,42 g/cm^3; SA10A (Diaion) – 1,15 g/cm^3; SK112 (Diaion) – 1,53 g/cm^3; SK116 (Diaion) – 1,51 g/cm^3.

Predetermined quantities of powders were mixed at ambient temperature in a small batch mill, and the mixture was loaded into a pre-heated kneader. The kneader Haake Bucher Rheocord "Fision 90" with double arm mixing chamber, exhibiting a mixing volume of 48 cm^3, was used. The powders were mixed for 20 min at 60 rpm. The set temperature of the kneader was 155°C for most of the mixtures, while only for mixtures consisting of elastomers with higher melt index the temperature was set to 110°C. Due to the friction energy dissipated during kneading, the measured temperature in the kneader is higher than the set temperature. For most of mixtures, like mixtures containing PE, the temperature in the kneader was between 170°C and 190°C.

The use of a kneader instead of an extruder, has the advantage that a number of different mixtures of resin powder and binder can be easier prepared and the most suited can be selected for the membrane production. At pilot or industrial scale the extrusion would be used preferably since it delivers the mixture in form of a flat sheet.

3.2.4   Pressed heterogeneous membranes

For the production of flat membranes a weighed portion of the kneaded mixture was loaded into a 16 cm × 16 cm press-mold. Fiberglass reinforced PTFE-films (Klaus Ehrig PTFE-Kunststoffwerk GmbH) were applied on both sides of the mixture to provide a good release of the membrane from the mold. The mass of mixture was estimated to produce a membrane of ca. 0,3 mm thickness. A press type P 300 P (Dr. Collin GmbH) was used for the pressing.

Pressing conditions used for most of tested mixtures are summarized in **Table 3.1**.

Table 3.1

Pressing conditions

stage	Press settings			
	$t,^{\circ}C$	τ, min	P, kg/cm^2 (weight on mold area)	process
1	160	10	0	heating
2	160	1	35	pressing
3	160	1	352	pressing
4	160	1	527	pressing
5	20	15	0	cooling

For mixtures containing elastomers with higher melting index a pressing temperature of 115°C was used instead of 160°C. After opening the press-mold a flat, semitransparent sheet of membrane is released from the PTFE films.

A specially designed press-mold was manufactured for pressing of profiled membranes. A photo of both parts of this mold is shown in **Fig. 3.7**. The notched area of the mold presents the negative of the profile shown in **Fig. 3.4*b***, exhibiting a number of parallel, 0,5 mm high notches of trapezoidal shape with a distance of 1,25 mm between the notches. The direction of the notches is 45° to the axis of the mold, and the notches of the upper and lower part of the mold are perpendicular to each other. The conditions listed in **Table 3.1** were applied for pressing of the profiled membranes.

Contrary to the flat membranes, a chemical release agent was sprayed over the mold surface to prevent a sticking of the membrane in the mold. EWO-mold 7901D (Eckert&Woelk GmbH) was selected as agent, which provided a good release of the profiled membrane from the mold.

To improve mechanical stability, two flat membranes were sandwiched together with an open mesh monofilament fabric PEEK 17-115×145/58 (Sefar Inc., Switzerland) between them and pressed into one membrane. The reinforced membrane shows significantly higher mechanical strength and lower expansion in length by swelling than the non-reinforced membranes made of the same material. In the swollen state the reinforced membrane shows a wave-like deformation of its planar shape. Since a high mechanical strength of membranes was not a main objective of this work, the further experiments were performed with non-reinforced membranes, which were generally planar, without deformations.

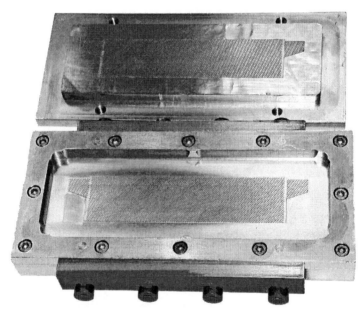

Fig. 3.7. Press-mold for the production of profiled membranes with a total area of 75 mm × 224 mm and a notched area of 40 mm × 160 mm

In some trials commercial heterogeneous reinforced ion-exchange membranes MK-40 (Shekinoazot, Russia) and Fumasep FTCM (FumaTech, Germany) were profiled in the press mold. One layer of such membranes is not thick enough to fill the press-mold and the profile was imprinted not satisfactory. Pressing two sheets of such membranes together resulted in a profiled membrane with a well developed profile, but the membrane produced was relatively thick since it contains four reinforcement fabrics.

3.2.4.1 Properties of pressed membranes

A line segment of 10 cm length was marked on a freshly pressed membrane. Thickness and mass of the membrane were measured. Then, the membrane was immersed into water for complete swelling and the same measurements were repeated. Swelling, conductivity and ionic transport numbers of produced flat membranes were determined for characterization of different mixtures of resin powder and binding polymer using the methods described in the Chapter 2.3.4.

In preliminary tests, strongly acidic resin Dowex C500 was mixed with ethylene-octene copolymer (Engage 8200 elastomer from the Dow Chemical Company) at different fractions. After swelling of membranes in water, their specific electric conductivity was measured.

Membranes containing volume fractions of ion-exchange resin $\varphi_{IER} < 0,60$ can be considered as non-conductive. At higher fractions a rapid increase of membrane conductivity with increasing φ_{IER} occurs. This behavior can be explained quantitatively by the percolation theory (see Chapter 2.3.1.3, **Fig. 2.13**). The membranes with $\varphi_{IER} = 0,64 - 0,68$ have a conductivity comparable to commercial heterogeneous ion-exchange membranes. Membranes with $\varphi_{IER} > 0,70$ turned out to be mechanically unstable. A volume fraction of $(67,5\pm1)\%$ of the ion-exchange resin in the mixture was therefore selected for the further membrane production.

A comparison of different polymers used as binder shows that the membranes produced from elastomers like Engage and Hytrel have a high expansion by swelling and weak mechanical stability in the swollen state. This makes them not very convenient for further use. Harder binding polymers and ion-exchange resins with lower swelling degree were therefore tested to improve mechanical stability and to prevent possible formation of larger gaps in the heterogeneous structure of the membrane by swelling.

PVC and polypropylene as binding polymers turned out to require temperatures which are higher than the thermal stability of ion-exchange resins. Trials with these polymers resulted in a strong smell during heating and mixing of powders and required high mechanical power for mixing, which was considered as not acceptable for further production. Moreover, membranes produced from these polymers exhibit a strong rigidity, even in swollen state and can easily be broken.

Therefore the following polyethylene (PE) products were tested as binding polymer: HDPE Stamylan HD 8621 (Sabic), LLDPE FG5190 (Borealis), LDPE Novex (BP), HDPE Lupolen 4261 AQ404 (BASF), HDPE Lupolen 1800S (Elenac). The two PE types mentioned first allow good mixing with the ion-exchange powder at 160°C. The mechanical properties of the produced membranes were found to be the best of all tested PE products. They were therefore used as binders for the final membrane production.

Different powders of strongly acidic or strongly basic resins based on sterene-divinylbenzene have been tested for membrane production. The resin powders tested are: Dowex C500 (H^+), Dowex C400 (Na^+), Finex C8100 (Na^+), Diaion-SK116 (Na^+), Diaion-SK112 (Na^+), Amberjet 4400 (Cl^-), and Diaion-SA10A(Cl^-). The test results show a difference in the ion-exchange capacity and in the swelling degree.

The swelling of a resin, the mechanical property of a binding polymer and its fraction in the mixture determine the swelling expansion of the produced membranes. The measurements show that for non-reinforced membranes the relative expansion in thickness and in length is the same within the accuracy of the measurements.

In **Fig. 3.8** the expansion by swelling of some membranes, produced from different resins and binding polymers at resin volume fraction 67,5vol.% are shown. **Fig. 3.8** indicates that for membranes produced from conventional resins with 8% of DVB, the swelling of cation-exchange membranes is stronger than for anion-exchange membranes. Thus, a membrane from CR Dionex C400 (Na^+) which has an ion-exchange capacity (*IEC*) of 2,2 eq/L_{bed} has a 1,5 times stronger expansion by swelling than a membrane with AR Amberjet 4400 (Cl^-) with an *IEC* of 1,4 eq/L_{bed}.

The use of a cation-exchange resin with a higher degree of cross-linking like 16% DVB for SK116, results in a membrane swelling similar to that of an anion-exchange membrane containing a resin with a lower cross-linking like 10% DVB for SA10A.

Since all membranes produced with 67,5vol.% of ion-exchange resin have good conductivity, the final mixture was chosen with respect to the mechanical properties and the swelling of the membrane. Since anion- and cation-exchange membranes should be produced in the same mold, the goal was to make membranes with identical dimensions in the wet state, which means with a similar degree of swelling.

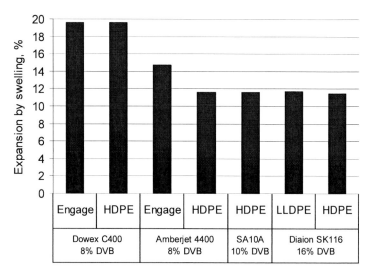

Fig. 3.8. Expansion by swelling in water of membranes made from different ion-exchange resins and binding polymers containing 67,5 vol.% of resin powder

3.2.4.2 Pressed membranes for electrodialysis tests

As resins Mitsubishi Chemical Diaion-SK166 (Na$^+$) and Diaion-SA10A (Cl$^-$) were selected for the membrane production. The grinding of the swollen and frozen resins resulted in powders with an average particle size 19 μm for the cation-exchange resin and of 35 μm for the anion-exchange resin.

The powders were dried and mixed with 32,5vol.% of HDPE Stamylan 8621 (Sabic) or LLDPE FG5190 (Borealis) in a kneader at 170°C (measured) and compressed into approximately 1 mm thick plates.

Pieces of these plates with predetermined mass were used for hot-pressing to a profiled membrane in the press-mold shown in **Fig. 3.7**, or to an approximately 0,3 mm thick flat membrane in the flat press-mold at conditions specified in **Table 3.1**. After pressing the membranes were immersed into water for complete swelling.

Both the cation- and the anion-exchange membranes prepared by the above described procedure expand about 11,5% by swelling in water. **Fig. 3.9** shows a photograph of the profiled membranes in the swollen and the dry state, showing the same dimensional changes.

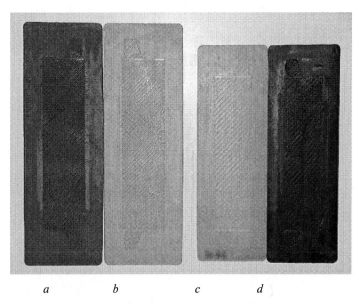

a   b   c   d

Fig. 3.9. Photograph of pressed profiled IEMs produced for ED-tests:
a - swollen in water CM; *b* - swollen in water AM; *c* - dry AM; *d* - dry CM

SEM images of the surface and the cross-section of flat membranes produced by pressing are shown in **Fig. 3.10**. It is evident that the membrane surfaces are mostly covered by the binding polymer. This is typical for heterogeneous type membranes. Comparing the cross-sectional views of the membranes produced in this work and of the commercially available heterogeneous membrane Fumasep FTAM (see **Fig. 2.14**) shows that the mean particle size of the here prepared membranes is somewhat larger than that of the FTAM.

Surface	Cross-section

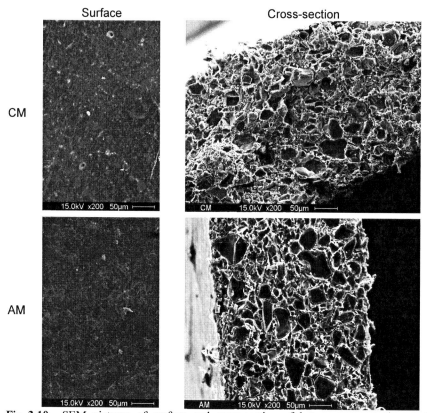

Fig. 3.10. SEM pictures of surface and cross-section of heterogeneous cation- and anion-exchange membranes produced by pressing and used for the tests

Conductivity, permselectivity and expansion by swelling were measured using the samples of flat membranes. The measured data are presented in **Table 3.2** and compared with data obtained for some commercial membranes. The comparison shows that the properties are rather similar.

Table 3.2

Properties of produced and commercial ion-exchange membranes

	Type	Binding polymer	IER	IER, vol.%	κ, S/m	t_{cou}*, %	Swelling in length, %	Notes
	AM	Stamylan HDPE 8621	SA10A	67,5	1,1	90	11,6	heterogeneous, non-reinforced
	CM	LLDPE FG5190	SK116	67,5	1,1	89	11,7	heterogeneous, non-reinforced
Commercial membranes		Manufacturer						
Nafion 117	CM	Du Pont	USA	100	2,53	98	-	homogeneous, non-reinforced
Fumasep FTCM	CM	FumaTech GmbH	Germany		1,5	92	1,7	heterogeneous, reinforced
MK-40	CM	JSC Schekinoazot	Russia	65	1,12	91	~8	heterogeneous, reinforced
Excellion I-100	CM	Electropure	USA	-	0,88	77	18,2	heterogeneous, non-reinforced
Fumasep FTAM	AM	FumaTech GmbH	Germany		1,7	88	1,8	heterogeneous, reinforced
Excellion I-200	AM	Electropure	USA	-	1,37	74	12,5	heterogeneous, non-reinforced
Ionac MA-3475L	AM	Sybron Chemicals	USA	-	0,44	89	-	heterogeneous, reinforced

*The transport number of counter-ions (t_{cou}) is measured potentiometrically between 0,5 m and 1,0 m KCl solutions (electromotive force measurements)

Photographs of produced membranes which are ready for assembling are shown in **Fig. 3.11**a and b. The key dimensions are indicated and for the notched membrane the location of two flow distribution/collection channels is marked. The shape and location of this flow distribution channel has been developed in flow visualization experiments described in the following section. After swelling of the pressed membranes in water, the outer shape and the openings for the inlets and the outlets were cut using a template.

a

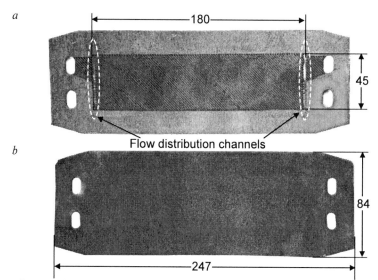

Flow distribution channels

b

Fig. 3.11. Photographs of membranes produced by pressing (dimensions in mm):
a – membrane with profile; *b* – flat membrane

The total area of the membrane swollen in water is 84 mm × 247 mm and it consists of a profiled area of 45 mm × 180 mm active for ionic transport and a non-profiled area for sealing with gaskets. Details of the shape and the dimensions of the profile are shown on the photographs of **Fig. 3.12***a* and *b*.

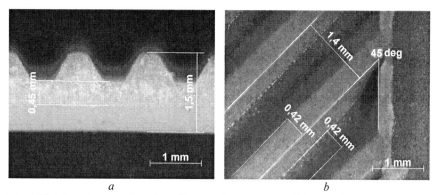

a   *b*

Fig. 3.12. Profile geometry of swollen membranes produced by pressing:
a – cross-section of an anion-exchange membrane; *b* – top-view of a cation-exchange membrane

3.2.4.3 Optimization of flow distribution

The flow through adjacent cell pairs in an electrodialysis stack is usually distributed through four collectors/distributors at the upper and lower ends of the stack. Two of them provide the feed and exit for the diluate and two for the concentrate. The ion-exchange membranes therefore contain four openings at their ends for the four channels as shown in **Fig. 3.11***a*. Here a feed may enter at the upper left opening, flows through the channels provided by the profiled section of the membranes and leaves the compartment at the upper right opening.

In order to fully utilize the whole active (profiled) membrane area, the flow should be evenly distributed over the width by a flow distribution and a flow collection channel situated at both ends of the flow field (see **Fig. 3.11***a*). The appropriate size and location of these channels has been determined in flow visualization experiments, depicted in **Fig. 3.13**.

For these experiments two plates, notched from one side like the membranes, were made of transparent epoxy-resin and assembled to resemble a flow channel between two profiled membranes. This channel was rinsed with water at velocities typical for electrodialysis. Blue ink was injected into the inlet stream and the resulting color patterns inside the compartment were recorded.

Fig. 3.13 shows sections of a video recorded at different times after ink injection. The color distribution indicates that, after a short initial distribution, a plug flow with a sharp front passes through the channel. It can be seen that the equal flow distribution results from the appropriate shape of the flow distribution channel depicted in **Fig. 3.11**.

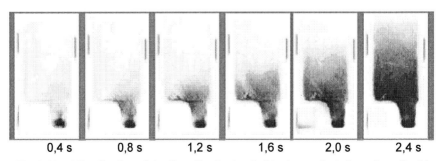

| 0,4 s | 0,8 s | 1,2 s | 1,6 s | 2,0 s | 2,4 s |

Fig. 3.13. Visualization of the flow distribution inside the notched channel, realized by ink-injection to the water stream of 2 cm/s

3.2.5   Calendered heterogeneous membranes

The production of profiled membranes by hot calendering of ion-exchange membrane sheets between corrugated rollers at elevated temperatures was also tested. The rollers with a triangular profile were produced by knurling of a brass roller using the Zeus knurling wheel (from Hommel + Keller GmbH). The picture of the knurling wheel, its pattern and the dimensions of the corrugated roller are shown in **Fig. 3.14** and **Fig. 3.15**. The surface profile of the corrugated roller has elevations of triangular form with an angle of 90°, a profile height of 0,5 mm, and 45° angle between the notches and the roller axis. The rollers have been used in a laboratory post-extrusion calendering system of Haake Buchler Instruments Inc.

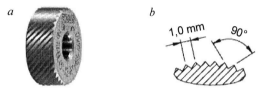

Fig. 3.14.   Knurling wheel (Zeus BL45 №10 / milled, with chamfer) used to profile a brass roller: *a* – picture of the wheel; *b* – dimensions of knurling wheel profile

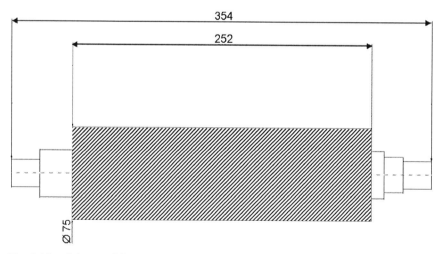

Fig. 3.15.   Scheme of the produced corrugated brass roller

Some flat heterogeneous membranes produced by pressing (as described in Chapter 3.2.3), as well as commercial heterogeneous membranes, were successfully profiled between the rollers. Self produced membranes turned out to have larger differences in thickness compared to commercial membranes. This required an individual adjustment of the gap distance between the rollers.

The membrane samples of pressed flat heterogeneous membranes are smaller (limited by the size of the used press-mold), while commercial membranes are available in long rolls. A longer streak of such membranes can be profiled between corrugated rollers in one batch and used in a larger membrane stack. Therefore, for the production of profiled membranes used in the subsequent tests, commercial heterogeneous ion-exchange membranes received from FumaTech GmbH (Germany) were used. Characteristics of these membranes are briefly summarized in **Table 3.2**. Every membrane contains two reinforcement fabrics on both sides, which provides mechanical strength and low stretching by swelling.

Just before entering the space between the rollers the membranes of FumaTech were heated up with a hot air stream to approximately 120°C, while the rollers were preheated to approximately 80°C. After passing between the corrugated rollers the membrane cools with ambient air. Thus profiled membranes were swollen in water and cut into pieces to the dimensions of the ED cell. The surface of the profiled membrane and its cross-section, perpendicular to notches of the profile is shown in **Fig. 3.16**.

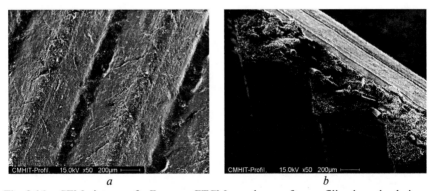

a   *b*

Fig. 3.16. SEM pictures of a Fumasep FTCM membrane after profiling by calendering: *a* – top view of notches; *b* – cross-sectional view of a sample broken in liquid nitrogen

3.2.6   Comparison of membrane production procedures

Both the pressing and the calendering techniques are applicable for the production of heterogeneous profiled membranes. In spite of a successfully produced series of pressed membranes the pressing procedure, created some problems which are listed below:

- the release of a membrane from the pressing-mold can damage the membrane (a reliable release requires the selection of the right release agent and a careful opening of the mold after pressing);
- in a number of cases the produced membranes showed some surface defects which appeared during the pressing (better drying of the components just before pressing and applying a vacuum during pressing reduced these surface defects, but did not solve the problem completely);
- the pressing in the mold takes a longer time (cycles of heating-pressing-cooling) than calendering;
- it was not possible to profile commercial membranes which were too thin for the used press-mold, and the development of thicker heterogeneous membranes was required.

The calendering procedure, on the other hand, has met significantly less obstacles and has proven to be more convenient and less time consuming than pressing. Considering the production of profiled membranes in pilot or industrial scale, calendering could be coupled with extruding, where the hot band of a flat heterogeneous membrane coming out of the extruder can be profiled between corrugated rollers.

3.3 Testing of the produced membranes in water desalination

Both pressed and calendered profiled ion-exchange membranes have been tested in comparison with flat membranes made of the same materials in an electrodialysis test stack under different operating conditions and with different feed water. The set-up of the stack and the test environment will be described in the following.

3.3.1 Stack flow scheme

In conventional electrodialysis the diluate and concentrate compartments are usually fed by solutions of identical composition. High concentration gradients across a membrane can arise, especially when the flow direction in both compartments is the same. Co-ions penetrating the membrane and reaching the diluate close to the inlet part of the diluate compartment can be removed during the further passage through this compartment, but for co-ions passing the membrane close to the diluate outlet, the remaining path length might be too short for their subsequent removal. If membranes with limited permselectivity like heterogeneous membranes are used, it is important to prevent a significant concentration gradient between concentrate and diluate in order to limit the transport of co-ions into the diluate close to the diluate outlet. Based upon preliminary experiments, a new flow scheme has been developed for the stack with profiled membranes in order to overcome the above mentioned limitations. In this flow concept (see **Fig. 3.17**), only the diluate compartment will be fed by the feed water. A part of desalted water from the diluate compartments leaves the stack as product, while another part is used to rinse the concentrate compartments in a direction opposite to the flow direction in the diluate.

In a conventional electrodialysis stack with flat membranes and non-conductive spacers, such a flow scheme would be questionable, since the current density at the diluate side would be very low due to the low conductivity of the diluate. Using profiled membranes, however, the conductivity is increased by the direct contact points of adjacent membranes.

Good conductivity of electrode compartments could be provided using a profiled, or corrugated, electrode being in contact with the last profiled membrane forming a flow through channel in a similar way as between two profiled membranes. Since profiled electrodes were not available and the investigation of electrode compartments was considered not as a main aim of this work, flat electrodes were used. The last membranes of the stack were pressed with profile only from one side and faced with the flat side to an electrode, while a neutral spacer was placed between a last membrane and an electrode.

Since the contact points are missing in the electrode compartments, the electrode compartments will be conventionally rinsed with water of higher ion conductivity, e.g.

with softened tap-water. In order to limit contamination of the diluate through the electrode rinse, a "protection compartment" has been placed next to each electrode, which is also flown through by part of the diluate produced (see **Fig. 3.17***a*).

The waste of the stack comprises the stream rinsing electrode compartments and the streams rinsing concentrate and protection compartments. The loss through the electrode and the protection compartments in the total recovery rate will be reducing with increasing number of repeating units in a stack. Moreover, as mentioned above, in case of a cross-corrugated structure between a membrane and an electrode in the electrode compartments, the use of protection compartments might not be required. In the following only the loss through the concentrate compartments will be considered in calculating the diluate recovery as ratio of the diluate product and the feed (Eq. (2.45)).

The flow scheme of **Fig. 3.17** has several advantages in preventing the penetration of contaminations from the concentrate into the diluate compartments. Since the pressure in a concentrate compartment is always lower than in a diluate compartment, a hydrodynamic cross-membrane transport of salts into the diluate compartment is impossible. Since the ion concentration of water at the outlet of a diluate compartment and at the inlet of a concentrate compartment is identical, the concentration difference between both membrane sides close to the diluate outlet is always small. This results in an ion-poor region at the diluate outlet (at the top of the stack in **Fig. 3.17**), where the transport of salts from concentrate into diluate is strongly reduced. In addition, the protection compartments next to the electrodes prevent the penetration of contaminants from the electrode rinse into the product water.

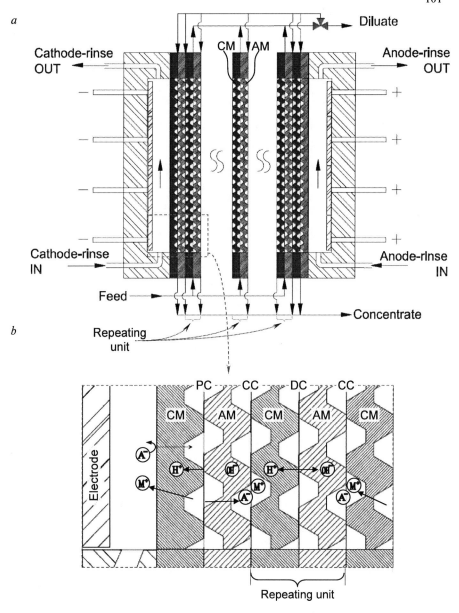

Fig. 3.17. Flow scheme of the stack used for the tests of ED with profiled membranes: *a* – general view; *b* – details of ionic transport. DC, CC and PC are diluate, concentrate and protection compartments; AM and CM are anion-exchange and cation-exchange membranes

3.3.2   Design of the test stack

For all subsequent experiments a stack made of 7 pairs of membranes was assembled according to the concept illustrated in **Fig. 3.17**, using flat or profiled membranes swollen in water. Similar membrane sizes as in **Fig. 3.11** were chosen for pressed, calendered and flat membranes.

In case of profiled membranes, the notches of one membrane stay in direct contact with notches of their neighbor membranes and are oriented perpendicular to each other. The overall structure of crossed notches provides a good mechanical stability of the stack. The sealing in the gasket parts between profiled membranes was achieved by using a two-component silicone rubber Elastosil® M 4540 with 5% of catalyst T 35 (both from Wacker Silicones Division). This silicone rubber was selected because of its ability to cover the surface of ion-exchange membranes swollen in water and because it solidifies at room temperature. For pressed membranes sealing was applied in the non-profiled parts of the membranes. Since calendered membranes are profiled over their entire surface, sealing was applied in about the same areas as in pressed membranes.

Both pressed and calendered membranes have been tested and compared with the respective non-profiled flat membranes. In the case of calendered ion-exchange membranes, Fumasep FTCM and FTAM (FumaTech GmbH) were used. In contrast to the pressed profile the calendered profile has notches of triangular shape and contains no flow distribution channels, like shown in **Fig. 3.11***a*.

In the desalination experiments with the self pressed flat membranes the woven monofilament open mesh fabric Propyltex 05-1000/45 (Sear Inc.) with a thickness of 1 mm was used as a spacer between the membranes. This mesh has the thickness close to that of notched channels and the mesh count of 6,8 filaments per centimeter, which corresponds to the distance between the notches of the pressed membrane profile. The spacer was cut and placed in the stack at an angle of 45° between the filaments and the main flow. Rubber gaskets with a thickness corresponding to the spacers were used.

For comparison with the stack build of calendered profiled membranes a second stack with flat Fumasep FTCM and FTAM membranes was assembled. The compartments formed by 0,63 mm thick rubber gaskets were equipped with the woven spacer Nitex 06-750/47 (Sefar Inc.). According to the specifications of the manufacturer this spacer has a filament diameter of 355 µm and a mesh count of 9 filaments per centimeter, which is about similar to the dimensions and the count of notches in the calendered profiled membranes.

The produced membrane stack was slightly compressed from the sides by end-plates made of Plexiglas, which are illustrated in **Fig. 3.18**.

The electrodes are positioned in the end-plates. The total dimensions of the electrodes and of the active part of the membranes in the assembled module is 45 mm × 180 mm. Both electrodes were divided into four segments with separate electrical connections to allow measurements of the current density distribution along the flow length. Conventional mesh screens were placed between electrodes and the last membranes to create the flow-through channels for rinsing the electrode compartments.

Collection and distribution of the feed, diluate and concentrate streams occur in the common collectors/distributors two of which are located at the top and two at the bottom of the stack, created by the holes in membrane (see **Fig. 3.11**). A flow divider at the exit of diluate splits it into a stream feeding the concentrate and the protection compartments and the stream of product water (**Fig. 3.18**b). Flow valves in the respective exit streams allow to control the amount of diluate which is used to purge the concentrate and the protection compartments, and thus to adjust the water recovery rate of the stack (Eq. (2.45)).

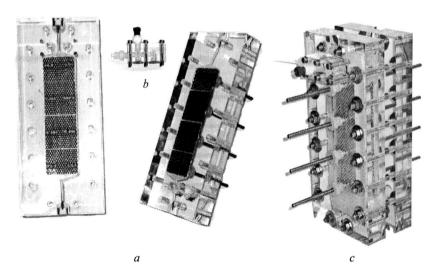

a   *c*

Fig. 3.18. Photographs of: a – end-plates with build-in electrodes; b – the flow-divider; c – assembled ED stack used for desalination tests

3.3.3   Set-up for desalination tests

Feed water from pretreatment block (**Fig. 3.19**) was provided to fill the tanks in the set-up for the tests of ED-stacks (**Fig. 3.20**). Tap-water is first treated by the softener ASW-120. The softened Stuttgart tap-water had a conductivity of about 330 µS/cm and a pH of 7,76 containing Na^+-ion as cation and different anions. Part of the produced softened tap-water (STW) (**Fig. 3.19**) is filled into the Tank 2 (**Fig. 3.20**) and used to rinse both electrode compartments of the ED stack.

The other part of the softened feed water is fed to the reverse osmosis unit RO-120 from BKG Wassertechnik GmbH (Germany) and the obtained RO-permeate, containing only Na^+-ions as mineral cation, is collected in a buffer tank, which has a volume of about 1 m^3. To maintain a constant conductivity of the RO-permeate produced, several modifications of the existing RO-unit have been made.

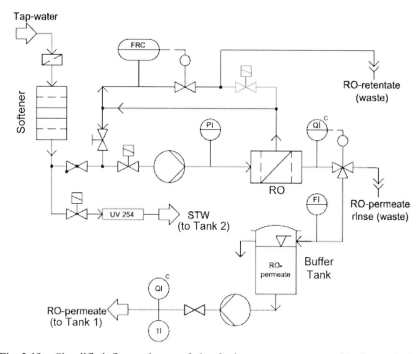

Fig. 3.19. Simplified flow scheme of the feed water pretreatment block, producing softened tap-water and RO-permeate, where QI^C relates to conductivity indicator, TI – temperature indicator, PI – pressure indicator, FI – flow indicator and FRC – flow controller

It is known that at the start of the RO operation after a long stand-by period the conductivity of the RO-permeate is initially relatively high, but decreases rapidly towards a stable value at steady-state. To prevent variations in RO-permeate conductivity, the RO-unit was operated non-stop as long as possible. When the buffer tank, collecting the RO-permeate, (**Fig. 3.19**) was almost full, the excess water was discharged through the side-overflow in the upper part of the tank. An Alicat Scientific™ LC Series Liquid Flow Controller was installed instead of the RO-retentate hand-valve in the RO set-up to keep the retentate flow rate and RO recovery rate constant in time.

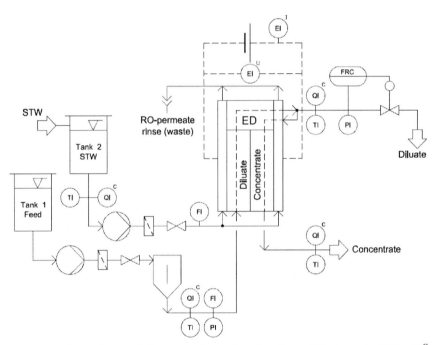

Fig. 3.20. Flow scheme of the set-up for testing the electrodialysis stack (ED). QI^C relates to conductivity indicator, TI – temperature indicator, PI – pressure indicator, FI – flow indicator, EI^I – current strength indicator, EI^U – voltage indicator and FRC – flow controller

During regeneration of the softener (a cation-exchange column in Na$^+$ form) the RO operation was automatically stopped for about two hours. Since the following restart resulted in a RO-permeate with a temporarily higher conductivity, an online measurement of the permeate conductivity after the RO-unit was installed. The conductivity cell was communicating with a relay. When the conductivity of the RO-permeate was higher than the set-value (ca. 10% higher than the typical conductivity for continuous RO operation) the relay switched the three-way valve in RO-permeate and the water with higher conductivity was discharged. Thus, only water with conductivity below the set-value was collected in the buffer tank.

Tank 1 and Tank 2 in **Fig. 3.20** have volumes of ca. 100 L.

For the following experiments different types of water were used to feed the diluate compartments of the ED stack:

- softened tap-water with a conductivity of 330 μS/cm;
- softened reverse osmosis (RO) permeate with a conductivity of 5,5 μS/cm;
- deionized water with a conductivity of 0,056 μS/cm, produced by the laboratory water purification system;
- 0,0005 M sodium chloride solution with a conductivity of ca. 63 μS/cm.

When softened tap-water was used as feed, the stack was connected to Tank 2. When RO-permeate was utilized as feed, Tank 1 was filled with RO-permeate produced in the pre-treatment block (see **Fig. 3.19**). When the NaCl solution was used as feed, Tank 1 was filled with deionized water under argon (Ar) atmosphere and a specified amount of crystalline NaCl was dissolved in it, while Ar was kept bubbling through the solution during the feed water preparation procedure and during the desalination experiments. In one set of experiments deionized water was fed directly to the stack without any intermediate storage tank.

In all experiments softened tap-water was used to rinse the electrode compartments.

Since the study was aimed to verify the possibility of obtaining low-conductivity diluate, relatively low water recovery rate was set. This should prevent obtaining highly concentrated concentrate and, together with rinsing of concentrate and protection compartments with a part of diluate in flow direction countercurrent to diluate, it should reduce co-ions leakage from concentrate to diluate. It needs to be taken into account that in such a small stack consisting of 5 diluate compartments 6 concentrate compartments and 2 protection compartments the contribution of protection compartments in recovery rate was not negligible. In most of experiments the flow rate through a concentrate or a protection compartment was a half of the flow rate through a diluate compartment. Thus, even at complete desalination, the concentration at the outlet of a concentrate compartment could not be higher than double of feed concentration and the overall recovery rate of the stack was 20%. In one set of desalination experiments higher recoveries were tested for comparison.

3.4 Experimental results and discussion

Results obtained from the desalination experiments with different current densities and flow rates and with different types of feed water, as specified above, are presented in the following sections. All results refer to steady-state conditions, which are usually reached after 20 min - 4 h of continuous operation. The values of the current density are calculated from the measured total current divided by the electrode area. The values of the potential drop over a membrane pair are calculated from the voltage drop of the whole stack divided by 7 (which is the number of cell pairs).

3.4.1 Desalination with pressed profiled membranes

Fig. 3.21 shows the diluate conductivity at different current densities and flow velocities in experiments with profiled membranes, utilizing softened tap-water and RO-permeate as feed. The desalination degree increases with decreasing diluate flow rate, i.e. increasing residence time, and with increasing current density. Interestingly, this trend does not continue for experiments with RO-permeate and flow velocities below about 3 cm/s. As can be seen for 2,4 cm/s diluate flow velocity in **Fig. 3.21**b, the diluate conductivity starts to slowly increase after a sharp drop to minimal conductivity, if the current density exceeds 1 A/m^2. At this value diluate conductivity is obviously so low that a further reduction of ion content by higher voltages (or current densities) is not possible. Instead, only the co-ions transport from the concentration and electrode compartments into the diluate is enhanced. This phenomenon will be further elucidated in Chapter 4.4.4.

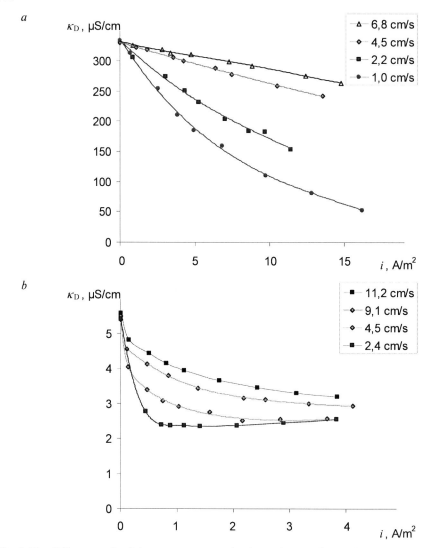

Fig. 3.21. Diluate conductivity versus current density and mean flow velocity through the diluate compartments of a module with pressed profiled membranes fed by: a – softened tap-water; b – RO-permeate; at diluate recovery of 20%

A comparison between the experiments with flat and with profiled membranes at a constant diluate flow velocity of 4,5 cm/s is given in **Fig. 3.22** to **Fig. 3.24**. In **Fig. 3.22** the diluate conductivity is shown as a function of the voltage drop over a cell pair with RO-permeate and softened tap-water as feed.

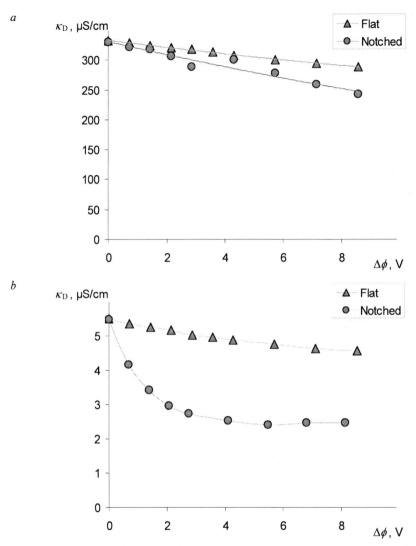

Fig. 3.22. Diluate conductivity as a function of the potential drop over a cell pair obtained in experiments with flat and profiled membranes at a flow velocity of 4,5 cm/s and diluate recovery of 20%, fed by: a – softened tap-water and b – RO-permeate

The results show that for desalination of softened tap-water the module with profiled membranes is not substantially superior to the one with flat membranes as long as the ionic conductivity of the solution is relatively high. If however the RO-permeate of significantly lower conductivity is used as a feed, the module with profiled membranes achieves substantially lower conductivities in the diluate than the module with flat membranes.

Fig. 3.23 shows the diluate conductivity as a function of the current density. The current-voltage curves for these experiments are presented in **Fig. 3.24**. They show higher current densities in the module with profiled membranes than with flat membranes at the same voltage. Again, the difference is more significant, when RO-permeate is used as feed and less pronounced when the softened tap-water is used.

If the module with flat membranes is fed by low-conductive RO-permeate the electrical resistance between the membranes in the diluate compartment is very high. Conversely, the electrical resistance is substantially reduced by profiled membranes due to the closer mean distance between the membranes and the direct contact points between adjacent membranes. This can clearly be seen from the current-voltage curves in **Fig. 3.24**b. For the highest possible voltage drop per unit cell of 9 V only a maximum current density of about 0,3 A/m^2 can be reached with flat membranes, whereas a more than ten-fold increase is possible with profiled membranes.

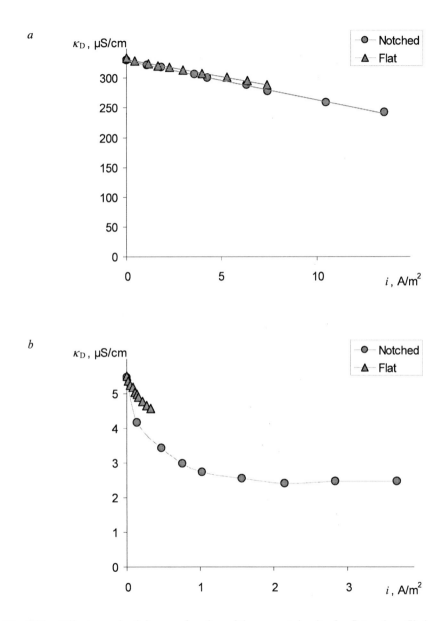

Fig. 3.23. Diluate conductivity as a function of the current density for flat and profiled membranes fed at a flow velocity of 4,5 cm/s by: *a* – softened tap-water and *b* – RO-permeate; at diluate recovery of 20%

a

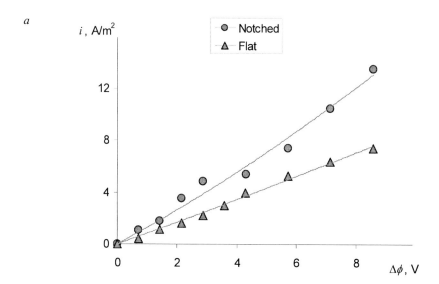

b

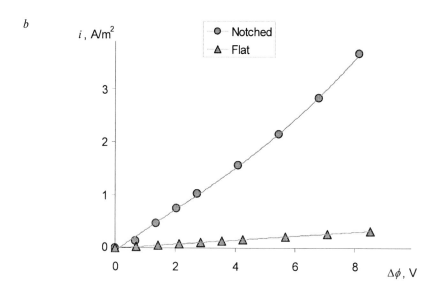

Fig. 3.24. Current density as a function of the potential drop over a membrane pair for flat and profiled membranes fed at a flow velocity of 4,5 cm/s by: *a* – softened tap-water and *b* – RO-permeate; at diluate recovery of 20%

3.4.2 Sequential batch experiments

In spite of better desalination with profiled membranes, the conductivity of the diluate produced with profiled membranes could not be reduced to the expected low level. In order to determine whether this was a result of a too short residence time in the electrodialysis stack, several subsequent batch experiments were performed. 100 L of RO-permeate were desalted in the first batch at a flow rate of 20 L/h through the module with profiled membranes at constant voltage of 15 V. The produced diluate (ca. 50 L) was collected and used as feed for the next batch at the same conditions of flow and voltage. A total sequence of seven subsequent batches has been performed until the diluate conductivity did not change any more. This simulates the performance of an ED stack with a very long process length.

The obtained data of the conductivity in the feed, the diluate and the concentrate as well as the current densities are summarized in **Fig. 3.25** for batch numbers 1 to 7. The results show that the difference between feed and diluate conductivity decreases with every next batch down to a minimum value of ca. 1 µS/cm, which could not be undercut. The conductivity in the concentrate is also decreasing with every next batch, but it is always significantly higher than in the feed.

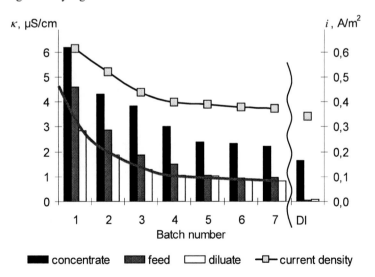

Fig. 3.25. Conductivity and current density in batchwise desalination of RO-permeate (κ_F = 4,6 µS/cm) with profiled membranes at a flow velocity of 4,5 cm/s (Q_F =20 L/h), diluate recovery of 50% (Q_C =10 L/h) and a voltage drop over the module of U_{cell}=15 V

The difference between the conductivities of the diluate and the concentrate is assumed to be caused by contaminations penetrating from the electrode compartments, which are always rinsed with softened tap-water with a substantially higher conductivity. This assumption will be tested below.

3.4.3 Experiments with deionized water as feed

The experiments with fully deionized water as feed were carried out to estimate the contamination of the diluate by co-ions penetrating from electrode compartments, which were always rinsed with softened tap-water. Deionized water was fed directly to the module with pressed profiled membranes. The flow rate of the feed water to the diluate compartments of the module was 20 L/h. 10 L/h of the water leaving the diluate compartments was used to rinse the concentrate and the protection compartments in countercurrent flow direction. The flow rate through the diluate compartments corresponds to an average linear velocity of 4,7 cm/s. Results of these experiments are presented in **Fig. 3.26** and **Fig. 3.27**.

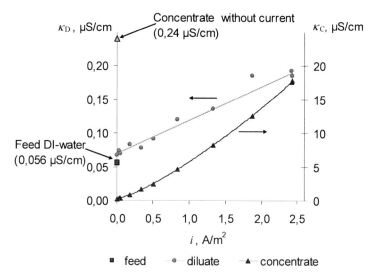

Fig. 3.26. Dependence of the water conductivity at the outlet of the diluate and concentrate compartments on the current density in the ED-stack with pressed profiled membranes fed by deionized water at flow velocity of 4,5 cm/s and diluate recovery 50%

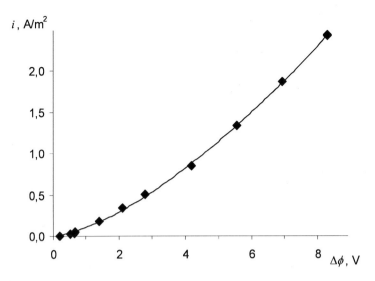

Fig. 3.27. Dependence of current density on potential drop over a cell pair in an ED-stack with pressed profiled membranes fed by deionized water

Fig. 3.26 shows an increase of conductivity in the outlet of the diluate (κ_D) and the concentrate (κ_C) compartments with increasing current density. The values of κ_D and κ_C are higher than the feed conductivity (κ_F), even if there is no current conducted through the cells. The increase of the water conductivity during the passage through the concentrate and protection compartments is significantly higher than the increase observed in the diluate compartments.

This increase of conductivity obviously occurs due to the migration of co-ions and the diffusion of electrolytes from the electrode compartments which are fed by water with a conductivity of 330 µS/cm. The water from the diluate compartments with a conductivity of 0,06 - 0,2 µS/cm is used to rinse the concentrate and protection compartments, which collect most electrolytes penetrating from the electrode compartments. Thus, most of the contaminants are rinsed with the concentrate before they can reach the diluate. This results in a significantly higher increase in the conductivity in the concentrate than in the diluate.

The increase of the water conductivity during the passage through the diluate compartments shows however, that the diluate compartments are not completely protected from contaminations due to the low permselectivity of the heterogeneous membranes. This is certainly a severe limitation which prevents obtaining ultrapure water by electrodialysis with profiled heterogeneous membranes.

The increase of conductivity with increasing current indicates, that contamination of concentrate and diluate is related to co-ion migration, whereas the effect of diffusion on conductivity is very low. It can be determined from measurements without applied current, where the conductivity in the concentrate reaches only 0,24 µS/cm.

The current-voltage curve in **Fig. 3.27** shows that, although no desalination occurs in the diluate compartments, a current is conducted through the membranes, which is probably due to the current provided mostly by H^+- and OH^--ions, generated by water dissociation at bipolar junctions between the membranes, as well as a certain current of co-ions. The current of H^+- and OH^--ions depends on the catalytic activity of the membranes to water dissociation and on the surface area of contacts between the membranes and it is possible, that this current could be different if membranes with another catalytic activity or another shape of profile are used.

3.4.4 Comparison of pressed and calendered profiled membranes

Experiments with calendered profiled membranes using softened tap-water as feed are shown in **Fig. 3.28** together with similar tests with pressed profiled membranes. The characteristics of profile geometry are summarized in **Table 3.3**. A low diluate flow velocity and a large recirculation ratio of diluate through the concentration and protection compartments (a low diluate recovery) was applied in order to obtain a high degree of desalination. It is evident from **Fig. 3.28** that the deionization performance of both types of produced heterogeneous membranes is quite similar. Only at low conductivities the profiled membranes prepared by calendering seem to perform somewhat better.

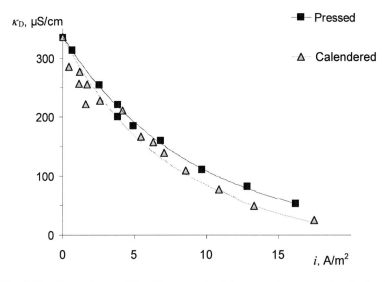

Fig. 3.28. Dependence of the diluate conductivity upon the current density in the stacks of pressed and calendered profiled membranes fed by softened tap-water at 1 cm/s flow velocity and 20% diluate recovery

Different reasons causing the somewhat better performance of calendered compare to pressed membranes could be a little larger membrane surface area contacting with water, shape of the notches providing easier ion-exchange between membrane and water, possibly the surface less covered by the binding polymer of the membrane, etc. Since the pressed and calendered membranes were made from different components using different techniques it is rather impossible to define unambiguously an explanation of the small difference in performance of tested membranes. The following experiments have therefore been performed with calendered profiled membranes.

Table 3.3

Characteristics of profiles in produced membranes

Production method	Pressed	Calendered
Shape of profile	∿∿∿	∧∧∧∧
Increase of surface area because of profile	42%	41%
Percentage of contact area to the total surface area	8-9%	< 1%
Cross-section area of the flow-through channel between profiled membranes, cm^2	0,23	0,23

3.4.5 Experiments with feed water free of low-dissociated compounds

In the previous tests only small differences in the performance between profiled and flat membranes were found with softened tap-water as feed. Utilizing RO-permeate as feed, the performance of the profiled membranes was significantly better than that of flat membranes, but even with profiled membranes a diluate conductivity of ca. 1 μS/cm could not be undercut. A possible cause could be the presence of weakly dissociated compounds, in particular of CO_2 taken up from the air. To test this assumption, a 0,0005 M NaCl solution was used as feed for the next experiments. This feed solution has a theoretical conductivity of 62,3 μS/cm, which is intermediate between typical tap-water and RO-permeate. To prevent a contamination of the solution with weakly ionized species, like CO_2, the NaCl solution was prepared by dissolving of NaCl in 100 L of deionized water ($\rho > 10$ MOhm·cm) under argon atmosphere and during experiments argon was bubbling through the solution. In this set of experiments a stack with calendered profiled membranes made from Fumasep FTCM and FTAM (see **Table 3.2**) is compared with a stack assembled with the original flat FTCM and FTAM membranes. In these desalination experiments a more powerful power supply than in experiments with pressed membranes is used, allowing operation at higher currents and voltages.

Fig. 3.29 presents the diluate conductivities measured in experiments with a flow velocity in the diluate compartment of 1 cm/s, while operation at different diluate recoveries was tested.

119

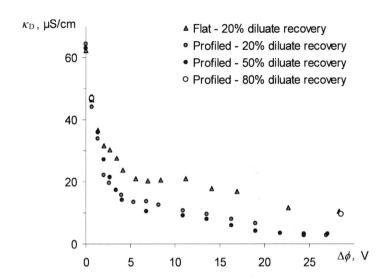

Fig. 3.29. Dependence of the diluate conductivity on the potential drop over a cell pair with calendered profiled membranes at flow velocity in the diluate compartment 1 cm/s

Fig. 3.29 shows that the diluate conductivity κ_D decreases with $\Delta\phi$, while a lower conductivity is reached in the stack with profiled membranes than with flat membranes at the same voltage drop and recovery rate. The relative difference in κ_D of both modules is larger at higher voltage drop. The decrease of diluate conductivity with increasing voltage drop has two main parts with different slops. When the diluate conductivity decreases down to ca. 20 µS/cm with flat membranes and to ca. 15 µS/cm with calendered membranes the further conductivity decrease occurs slower.

The lowest diluate conductivity achieved with calendered profiled membranes is ca. 2 - 3 µS/cm. This is comparable to the lowest conductivities reached in single pass experiments with RO feed (e.g. **Fig. 3.24**) and leads to the conclusion that the presence of weakly dissociated electrolytes is not the sole reason for the limited desalting ability.

Desalination performance at diluate recoveries 20% and 50% is quite the same, while the increase of diluate recovery to 80% at high voltage drop results in a higher diluate conductivity. This can be caused by an increase of the concentration in the concentrate stream and consequent increase of co-ions leakage from concentrate into diluate, which is remarkable at such low conductivities.

The current-voltage curves of this experimental series are shown in the **Fig. 3.30**. Similar to the tests with pressed membranes the stack with profiled membranes has higher current densities at a given voltage drop than the stack with flat membranes. It is due to the lower electric resistance of the contacting profiled membranes and it is resulting in better ions removal. Also the water dissociation at the bipolar junction between membranes, leading to H^+- and OH^--ions presence in the corresponding membranes, could reduce the electrical resistance additionally and remove more ions because of an additional ion-exchange between the regenerated membrane and the solution.

Samples of the diluate were analyzed as shown in **Fig. 3.31** together with conductivity values. The concentration of Na^+-ions was measured by atomic absorption spectrophotometry (AAS) and that of Cl^--ions by ion-chromatography (IC).

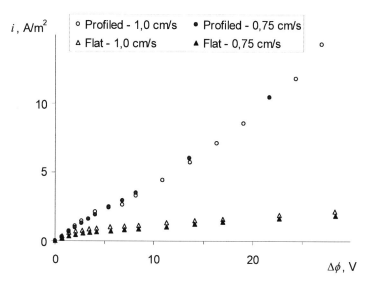

Fig. 3.30. Dependence of the current density on the potential drop over a cell pair with calendered profiled membranes at diluate flow velocity 1 cm/s and 20% diluate recovery

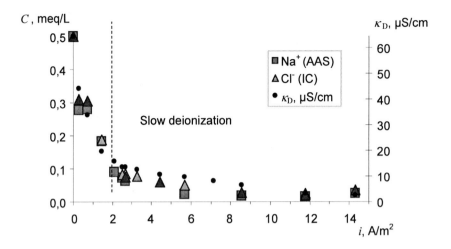

Fig. 3.31. Dependence of Na$^+$- and Cl$^-$-ion concentrations and conductivity in the diluate on the current density in the module with calendered profiled membranes fed by 0,0005 M NaCl solution at 1 cm/s flow velocity and 20% diluate recovery

Both Na$^+$- and Cl$^-$-ions have equivalent concentrations, which are considered as equal within the measurement error. The decrease of κ_D or of the ionic concentrations of the diluate could be separated into two parts: 1) a "high-efficient" desalination region at low current density and 2) a "low-efficient" desalination region at high current densities.

In the „highly-efficient" region in **Fig. 3.31** a strong decrease of the diluate conductivity with the current is observed up to current densities of ca. 2 A/m^2, which corresponds to the concentration of salt ions down to 0,1-0,2 meq/L. The current efficiency in this region is higher than 50%.

In the "low-efficient" region only a slow decrease of κ_D with i takes place. Obviously a big part of the current is used for the water dissociation and for the "circulation" of ions between the concentrate and diluate compartments (co-ions penetrating through the membranes). The current efficiency in this region decreases steadily with the current density. At the highest current densities a small increase of κ_D with i can be observed similar to that at the lowest flow velocity in **Fig. 3.21**b.

The current efficiency for ion removal was calculated from the analysis of the ions and is shown in the **Fig. 3.32** together with the energy consumption for the production of 1 L of diluate.

122

Fig. 3.32 illustrates that in the "high-efficient" region of current densities the energy consumption is relatively low and the application of desalination with profiled membranes could be economically favorable. In the "low-efficient" region the economics of the operation become questionable for practical applications since only a small additional reduction of conductivity requires large additional energy consumption.

Fig. 3.33 compares the desalination with calendered profiled membranes at two different feed flow velocities and the current-voltage curves for these experiments are present in Fig. 3.30. It shows that lower diluate conductivities can be achieved at lower flow velocity (longer residence time). The lowest achieved value of κ_D is 1,4 μS/cm at a flow velocity of 0,75 cm/s, which is already close to the lowest conductivity obtained with pressed membranes in subsequent batch experiments (Fig. 3.25).

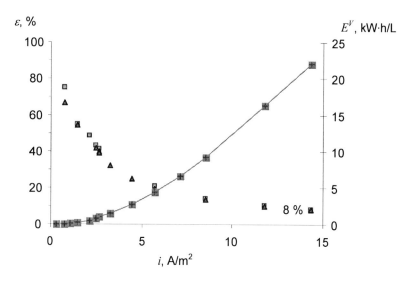

Fig. 3.32. Dependence of the current efficiency and energy consumptions upon the current density in the module with calendered profiled membranes fed by 0,0005 M NaCl solution at 1 cm/s flow velocity and 20% diluate recovery

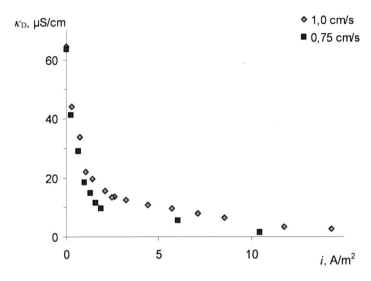

Fig. 3.33. Dependence of the diluate conductivity upon the current density in the module with calendered profiled membranes fed by 0,0005 M NaCl solution at 20% diluate recovery

3.5 Summary and outlook of electrodialysis with profiled membranes

The production of membranes profiled from both sides, using the corrugated rollers, is found to be the most simple, fast and efficient method tested. The results of the experimental study show definite advantages of an ED-stack with profiled membranes over one with flat membranes, as expected from theoretical considerations.

The stack with profiled membranes has lower electrical resistance and thus, lower energy consumption for the desalination process. The reduction in the energy consumption is especially significant when feed water of relatively low conductivity is processed. This extends the application range of electrodialysis to lower conductive solutions, making it more economic and ecologically sustainable.

Other important features of the profiled membranes are: closer spacing between membranes without the need of (expensive) spacers, no "shadow effect" on the electric field by spacers, increased membrane surface area and good turbulence promotion.

Although there is a possibility of electrochemically enhanced water dissociation at the contact points between membranes, there is no pH change, since the produced ions are permeating directly into the corresponding membranes and recombine to water in the concentrate. Furthermore, the generated H^+- and OH^--ions can regenerate the corresponding membranes into the H^+-ions or OH^--ions form, which could positively affect the ion-exchange between the membrane and mineral ions from water. However, the ion-exchange does not seem to be significant with the profiled membranes tested. This is probably due to the fact that the surface of the heterogeneous membranes is covered to a high extent by the binding polymer, which is not available for ion-exchange as indicated in the pictures of **Fig. 2.14** and **Fig. 3.10**. Ion-exchange resin particles are only visible on a small part of the surface and will not provide such an intensive ion-exchange as ion-exchange resin beads in conventional ion-exchange beds do. Therefore, most of generated electrochemically H^+- and OH^--ions just migrate through the corresponding membranes and recombine in concentrate compartments.

In spite of the above listed advantages, the lowest diluate conductivity achieved in experiments is far above that of ultrapure water. The use of relatively high conductive water in the electrode compartments together with the use of heterogeneous membranes with low permselectivity creates certain limitations for the achievable diluate conductivity. This problem could be overcome by using ion-exchange membranes of better permselectivity as well as by improvements of the electrode compartments which would allow rinsing the electrodes with water of significantly lower conductivity, e.g. RO-permeate, or diluate. This can be achieved if the electrode has a direct contact with the adjacent membrane. To provide flow through the electrode compartment, the electrode should be formed as a flow-through mesh or it could be a corrugated plate, e.g. with a notched surface, so that the electrode and the last membrane form channels similar to those between the notched membranes in the stack. To prevent profiled membrane which is in contact with profiled electrode from degradation by products of electrode reactions, such an electrode compartment can play a role of protection compartment and be rinsed with a part of the diluate.

Another restriction for complete deionization with heterogeneous profiled membranes is the difficulty to remove weakly dissociated electrolytes. As far as these contaminations are present in non-dissociated form, they are not impacted by the electric field and may diffuse through the membranes according to their concentration gradients.

An ion-exchange with a resin in the H^+-ions or OH^--ions form as used in CEDI allows to remove weakly dissociated electrolytes much more efficiently, but would require homogeneous profiled membranes as discussed above. A development of homogeneous profiled membranes with high ion-exchange capacity could therefore improve the ability of ions removal and produce a diluate of even lower conductivity. Optimization of flow channel geometry, especially the shape and dimensions of notches might also further improve the desalination performance.

One interesting direction of investigations, which could be of importance for electrodialysis with profiled heterogeneous membranes is the orientation of the ion-exchange particles inside the binding polymer of the membrane, improving conductivity over membrane thickness. It is shown in [132,133] that strong electric field of alternating direction during membranes formation can align the ion-exchange particles into chains, which substantially reduce the percolation threshold and increase the ionic conductivity through the membrane. Applying this procedure to the here discussed heterogeneous membranes requires heating of the membrane above the glass transition point for reorientation of ion-exchange resin particles in the applied electric field. This could be combined with the profile forming of the hot membrane surface by pressing or calendering.

Ideally the procedure of membrane profiling should be carried out at the side of a membrane manufacturer where the use of e.g. corrugated rollers could be easily implemented into the manufacturing chain.

4 Continuous Electrodeionization (CEDI)

Continuous electrodeionization is a combination of ion-exchange and electromembrane desalination. In electrodeionization the removal of ions from a solution by an ion-exchange resin takes place simultaneously with the electrochemical regeneration of the resin by H^+- and OH^--ions, produced by electrochemical water splitting. As a result, a non-interrupted removal of ions from the feed water occurs, providing the possibility for a continuous desalination.

The design of a typical CEDI stack is principally similar to that of an ED stack. It consists of a series of unit cells, separated by ion-exchange membranes and arranged between two electrodes. The diluate compartments of a CEDI stack are usually thicker (2-20 mm) than those of an ED stack and are filled with a bed of ion-exchange resin within specially formed frames, which provide sealing and flow distribution. The arrangement of membranes and resins in the stack differs depending on the process concept. In the following the different currently known concepts of CEDI will be briefly reviewed. One of the most promising concepts will be studied concerning its present shortcomings and possible improvements.

4.1 Previous developments in electrodeionization

In the first electrodeionization systems used for water desalination a mixed-bed ion-exchange resin was placed in the diluate compartment of an ED-stack, as proposed 1955 by W.R. Walters et al. [134] and shown in **Fig. 2.22**. This concept was later specified in several patents, e.g. [135-138]. The principles of CEDI with mixed-beds were discussed in more detail by E. Glückauf [139].

Pioneering experimental studies of CEDI [134,139,140] aimed at the treatment of dilute radioactive waste and a first pilot CEDI module was constructed by the Permutit Company (UK) for the Harwell Atomic Energy Authority at the end of 1950th [141,142].

The utilization of electromembrane processes with filled diluate compartments was also proposed and investigated for the removal of dissolved oxygen, and ionizable gases, e.g. CO_2, NH_3, etc. [143,144], as well as for the production of high purity water [145-148]. However, only more than 30 years after its first description CEDI became commercialized by Millipore Corporation [149,150].

Meanwhile, CEDI for the production of high purity process water is a cost effective alternative to ion-exchange desalination [151]. Advantages of CEDI compared to conventional ion-exchange are its continuous operation and costs savings since no ion-exchange resin regeneration is required [152,153]. CEDI systems are now operating on different scales and applications with capacities up to a few hundred cubic meters product water per hour [142].

CEDI usually uses a feed water with a conductivity of 2 - 40 µS/cm (more often 5 - 15 µS/cm) and yields a diluate with a conductivity of 0,06 - 0,2 µS/cm. In most applications for water deionization CEDI is coupled with reverse osmosis as pre-treatment. Depending on system requirements and feed water quality, CEDI is frequently combined additionally with filtration, water softening, oxidation and sterilization as a pre- and post-treatment, and a mixed-bed ion-exchanger as a final polishing step.

CEDI systems have found applications in industries with the most demanding high purity water requirements, such as in the manufacturing of semiconductors and pharmaceuticals, in analytical laboratories, in the surface finishing of electronic components, in the production of high quality optics and water for high pressure boilers used for power generation. CEDI can also be applied for the removal and recovery of radioactive, heavy and noble metal ions from industrial effluents [154-157]. The use of CEDI in a membrane reactor has also been studied [158-160].

Around 300 patent applications concerning CEDI have been published by now, mostly during the last two decades. This reflects an extending interest to application and commercialization of CEDI. The growth of the pure and ultrapure water market and a trend to replace ion-exchange deionizers by CEDI modules has initiated the formation of an increasing number of companies supplying this market with CEDI-modules or complete water treatment systems.

Instead of granular ion-exchange resins, in several publications the use of woven and non-woven fibers [156,161-165], foams [166,167], or porous ion-exchange monoliths [168,169] as conductive spacers is described. But since granular ion-exchange resins are produced in large amounts at relatively low prices (in the range 1-10 EUR/L for standard grade resin), they are still used in most CEDI-stacks.

4.2 Basics of CEDI with mixed-bed ion-exchange resin

The first CEDI stacks have been constructed similar to a conventional ED stack with cation- and anion-exchange membranes in alternating series, forming individual compartments (**Fig. 2.22**). Repeating cell pairs in which the diluate compartments are filled with a mixed-bed ion-exchange resin and the concentrate compartments with a non-conductive spacer are arranged between the electrodes. This stack concept, applied by Millipore [149,150], Ionics [188], Ionpure [170] and others was a prototype for the further developments.

The hydrodynamics of the water flow inside a CEDI module and its distribution over the different compartments belong to the most important design aspects which affect the overall efficiency of the stack. The pressure distribution inside a module is also important. Poor pressure distribution can cause leaks, deformation of membranes and the formation of voids which can result in poor current distribution.

Two mechanisms of ion removal from a feed water into the ion-exchange material can be distinguished. The first mechanism is an enhancement of ions uptake compared to conventional ED, where ions from the solution migrate under the influence of an electrical field directly into the respective ion-exchange membrane and further into the concentrate compartment. Additionally ions can also be taken up by ion-exchange beads and migrate through a continuous sequence of ion-exchange beads of the same polarity into the respective membrane and further into the concentrate. Such a sequence of ion-exchange beads can be considered as an extension of the respective membrane surface through which the effective membrane surface area is significantly increased. This enhancement effect is basically similar to the effect exploited with profiled membranes in Chapter 3.

The amount of ions removed by the enhancement effect depends on the ratio of the current passing through the water phase to that passing through the ion-exchange bed. As shown in the Chapter 2.2.4 the conductivity of a standard cation-exchange resins in equilibrium with solutions containing cations with valence $+1$ or $+2$ is equal to the conductivity of such solutions with concentrations of $(0,1 - 1)$ eq/L. Since RO-permeate, which is typically used as feed for CEDI, has a conductivity $2 - 5$ orders of magnitude lower, the current will be mostly conducted through the ion-exchange resin and only a negligible portion of the current will pass through the water between the beads.

130

The second mechanism of ions removal is caused by ion-exchange and the subsequent electrochemical regeneration of the ion-exchange resin by H^+- and OH^--ions, generated by water splitting. The mineral cations (Me^+) and anions (A^-) from the feed are exchanged similar to Eqs. (2.34) and (2.35) by H^+-ions from the cation and OH^--ions from the anion-exchange resin according to the following reactions:

$$\overline{CR^-H^+} + Me^+ \rightarrow \overline{CR^-Me^+} + H^+,$$

$$\overline{AR^+OH^-} + A^- \rightarrow \overline{AR^+A^-} + OH^-, \qquad (4.1)$$

where $\overline{CR^-}$ indicates fixed ions of CR, and $\overline{AR^+}$ indicates fixed ions of AR.

In a mixed- bed of ion-exchanger beads the ion-exchange equilibrium of both reactions in Eq. (4.1) is shifted to the right, due to the recombination of H^+- and OH^--ions, giving water:

$$OH^- + H^+ \rightarrow H_2O. \qquad (4.2)$$

The kinetics of ion-exchange in dilute solutions is governed by the film diffusion as mentioned in Chapter 2.2.

Electrochemically enhanced water dissociation can occur at the bipolar interfaces of the ion-exchange materials in the diluate compartment [139,145], as shown schematically in **Fig. 4.1**. In dilute solutions its intensity depends on the strength of the electric field, the area of bipolar contact and the type and concentration of fixed groups, which have a catalytic activity for the water dissociation reaction [171,79,80]. This means that one resin may be regenerated better than a counter-polar one.

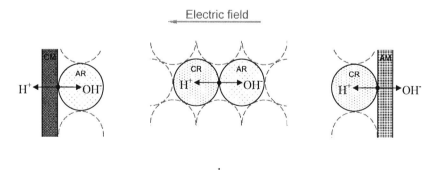

a   *b*   *c*

Fig. 4.1. The bipolar junctions between ion-exchange materials where electrically induced water dissociation occurs: *a* – cation-exchange membrane (CM)/anion-exchange resin (AR); *b* – cation-exchange resin (CR)/AR; *c* – CR/anion-exchange membrane (AM)

In a diluate compartment containing a mixed-bed, enhanced water dissociation will take place inside the mixed-bed at the bipolar junctions as shown in **Fig. 4.1** or **Fig. 4.2**a. The regenerated ion-exchange beads can subsequently take up cations (Me^+) or anions (A^-) from the solution. If such a bead is connected with the respective membrane by a sequence of beads of the same polarity the ions will migrate into the respective concentrate.

If this is not the case, a so called reverse junction occurs as in **Fig. 4.2**b, where the ions are forced back into the solution. In addition to the case shown in **Fig. 4.2**b where A^--anions and Me^+-cations are released into the solution as a salt, it is more likely that in the diluate A^--anion meets H^+-ion at a reverse junction, forming an acid, or Me^+-cation meets OH^--ion, forming a base. All the phenomena taking place at the reverse junctions, including the recombination of H^+- and OH^--ions, reduce the efficiency of the ion removal.

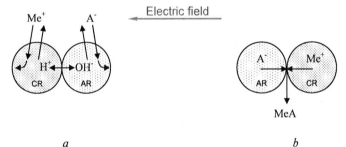

a   *b*

Fig. 4.2. Orientation of counter-polar beads of ion-exchange resins in the electric field of a CEDI unit: *a* – bipolar junctions; *b* – reverse junctions

Due to the random distribution of cation- and anion-exchange beads in a mixed bed, the average number of reverse junctions per membrane area depends on the ratio between the diameters of the beads and the thickness of the diluate compartment. In channels with a monolayer of mixed beads no reverse junctions can occur. With an increase of the compartment thickness the number of reverse junctions will increase and the current efficiency will be decreased. Hence, relatively thin compartments are used in CEDI with mixed-beds in order to minimize the number of reverse junctions and to have a lower voltage drop over a cell pair.

Under optimized conditions of current and flow rate a complete deionization of strongly dissociated electrolytes can be carried out successfully in a mixed-bed CEDI stack. In some cases the removal of mineral ions down to less than 1 ppb or even to levels not detectable with modern analytical techniques is possible [172]. On the contrary, the complete removal of weakly dissociated substances, which in practical applications usually relates to the silica and boron content, remains a severe problem and is hard to achieve with a mixed-bed in the diluate compartment.

Usually, suppliers of mixed-bed CEDI claim maximal silica content in the diluate of $2 - 10$ ppb. In [173] the feed water with 1000 ppb of SiO_2 and 50 ppb of boron was deionized by mixed-bed CEDI and in the best case the concentrations of SiO_2 in the product water was reduced down to 2,7 ppb and 1 ppb, respectively (without specification of experimental conditions). Such values are too high to meet the requirements for ultrapure water (see Appendix 3) and a downstream polishing step with additional ion-exchange columns is required for ultrapure water production.

4.2.1 Leakage of co-ions

As discussed in the Chapter 2.4.4, an ion-exchange membrane does not exclude the co-ions completely. Therefore they can penetrate the membrane to some extend. Under the driving force of an electrical field the co-ions will be transported in the opposite direction to the counter-ions, e.g. from the concentrate to the diluate compartment. This limits the product quality and reduces the process efficiency.

The effect of co-ion penetration is especially pronounced, when the concentrate compartment is not filled with ion-exchange material. Then the current is mostly transported through the *solution/membrane* boundary, resulting in an increase of the electrolyte concentration close to membrane surface facing the CC due to concentration polarization (Eq. (2.41)).

In the CEDI with mixed-bed both types of co-ions, penetrating the anion- and the cation-exchange membrane, can be present in the diluate as salt. But since usually the transport numbers of co-ions through cation- and anion-exchange membranes are different, either an excess of cations can be compensated by OH^--ions, forming a hydroxide, or an excess of anions can be compensated by H^+-ions forming an acid.

In general, the current density at the entrance and the exit of the diluate compartment has an opposite influence on the conductivity:

- in the entrance part an increase of the current density increases the regeneration degree of the ion-exchange resins in the diluate compartment, resulting in good ion removal and a substantial decrease of diluate conductivity;
- in the exit part of the diluate compartment, where the ions are almost completely removed, an increase in the current density results in a higher amount of co-ions transported into the diluate and in an increase of the diluate conductivity.

This interaction between voltage and current at the diluate compartment entrance and exit could result in a minimum of the diluate conductivity over current density as well as over diluate flow velocity, as observed by Pevnitskaya M.V. et al. [174]. They studied CEDI with a mixed-bed and a concentrate compartment filled with a neutral spacer and rinsed with NaCl solution at variable current I and different flow velocities v. Some results of this work are represented in **Fig. 4.3**, where a minimum in the $\kappa_D - I$ and κ_D - v dependencies is observed.

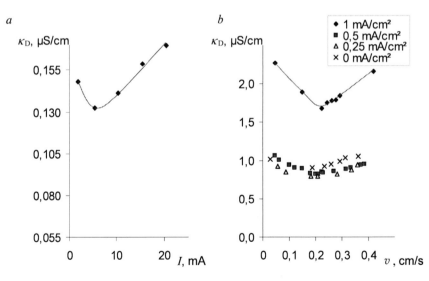

Fig. 4.3. Experimentally observed variation of diluate conductivity κ_D with current i in a mixed-bed CEDI from [174]: a – dependence of κ_D upon current, measured 90 min after start of operation at $v = 0,2$ cm/s with feed conductivity $\kappa_F = 5,6$ µS/cm and $t = 25$°C; b – dependence of κ_D upon flow velocity v at different current densities at $\kappa_F = 50$ µS/cm and $t = 30$°C

4.3 CEDI with layers of anion- and cation-exchange resin beds

Already in one of the first inventions concerning CEDI [136] not only a mixed-bed, but also either continuous layers of cation- and anion-exchange resins parallel to the flow direction, or subsequent layers perpendicular to the flow direction were claimed for the diluate compartment. Such arrangements of ion-exchange resins avoid reverse junctions in the diluate compartments of CEDI.

To distinguish between the two above mentioned orientations, a bed of a resin oriented parallel to the flow will in the following be called "stratum", like in the original patent [136] from 1957, while a bed oriented perpendicular to the flow will be called "layered bed", like for some commercialized CEDI modules [196].

4.3.1   Strata parallel to the flow direction

The formation of reverse junctions in a thin diluate compartment can either be avoided with a mono-stratum (compartment thickness of one bead) of mixed ion-exchange resin beads, or with ordered bi-strata consisting of one stratum of cation-exchange resin beads, adjacent to the cation-exchange membrane and one stratum of anion-exchange resin beads, adjacent to the anion-exchange membrane [136] as shown in **Fig. 4.4a**. Then each stratum can be considered as an extension of the respective membrane surface area for uptake of ions. In the electrochemical regeneration regime the electric potential gradient will induce water dissociation at the bipolar junctions between a resin and a membrane (**Fig. 4.1a** and c) in case of a mono-stratum and between the strata of the anion- and cation-exchange resins (**Fig. 4.1b**) in case of a bi-strata and the resins are then regenerated with H^+- and OH^--ions. Since the water flowing through both parallel strata will be well mixed, the removal of both, cations and anions, takes place.

Bi-strata can also be thicker than the size of two beads [136], as shown schematically in **Fig. 4.4b**. But such strata of two different ion-exchange resins should be relatively thin, so that sufficient lateral mixing of water streams across the thickness of diluate compartment occurs. Otherwise, the deionization efficiency decreases and in addition the increase of pH around the anion-exchange beads in OH^--ion form could cause scaling, if multivalent ions are present. Also the hydrodynamic conditions, determined by bead size and packing density of the different resins, must be balanced to prevent different flow rates through both strata.

The arrangement illustrated in **Fig. 4.4** allows for equal current passing through the anion- and cation-exchange bead strata, avoiding reverse junctions. But an additional stabilization of the strata by spacers or special devices is required to keep them fixed during operation.

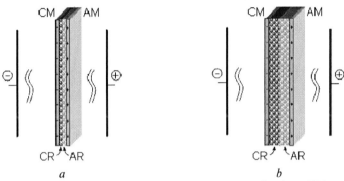

a   *b*

Fig. 4.4.   Diluate compartment filled with continuous strata of cation- (CR) and anion-exchange resin beads (AR) oriented towards the respective cation- (CM) or anion-exchange membranes (AM): *a* – ordered bi-strata in a thin compartment; *b* – ordered strata of CR and AR building thicker diluate compartment

4.3.2   Layers and clusters perpendicular to the flow direction

Other orientations of partially separated cation- and anion-exchange resin beds in the diluate compartment, which refer to so-called clustered beds [175] and alternating layered beds [136,176] are schematically shown in **Fig. 4.5**. Such systems have the diluate compartment filled with clusters alternating along the flow path clusters or layers of cation- and anion-exchange resin, extending over the whole thickness of the diluate compartment. Here the water dissociation occurs at the bipolar contacts between ion-exchange resins and membranes of counter polarity [177] (see **Fig. 4.1***a* and *c*). The clusters or layers of ion-exchange resins provide a migration path for regenerating ions through the resin from one membrane to the other and keep the resins regenerated and able for ion-exchange with mineral ions from the feed.

The idea of layered beds in CEDI was already described in patents in the 1950's [136] and 1980's [178]. The practical application of both, clustered, as well as

136

layered beds for water deionization by CEDI was discussed in the second half of the 1990's. Presently CEDI with clustered and layered beds is commercially applied in water deionization [204,179]. Compared to CEDI with mixed-bed, clustered or layered beds decrease the electrical resistance of the diluate compartment and improve the deionization performance, for the removal of weakly dissociated electrolytes. As reported in [180] a diluate resistivity of about 18 MOhm·cm and over 99,9% silica removal can be achieved with a CEDI stack using alternating layered beds.

Drawbacks of the CEDI with clustered or layered beds can result from the different conductance of anion- and cation-exchange resin beds and from the different intensity of water dissociation at the boundaries between the resins and the respective membranes. This can lead to local differences in the current passing through the bed which lead to differences in the regeneration of both resins. CEDI with layered beds is more sensitive to the leakage of co-ions from concentrate compare to CEDI with mixed-bed. In order to prevent co-ions leakage, in [176] a part of diluate is utilized to rinse compartments adjacent to the diluate compartment with layered beds. Arrangement of layers or clusters requires special equipment for mechanical [175], or electro-magnetic [181-183] layering of the resin in the diluate compartments of the CEDI stack.

In mixed-bed CEDI, due to the limit of the compartment thickness to few beads, the particle size should not be small. This limitation does not exist in a clustered or layered bed. Here thicker compartments can be used and the main parameter limiting the size of the beads is the pressure drop in the compartments.

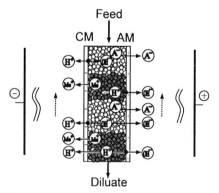

Fig. 4.5. Scheme of ion transport and electrochemically enhanced water dissociation in alternating layered beds between cation-exchange (CM) and anion-exchange membranes (AM) in a diluate compartment of CEDI

4.4   CEDI with separated beds

The problems of reverse junctions in mixed-beds and the problems of unequal flow or current distribution through the cation- and anion-exchange resins in clustered or layered beds can be overcome by CEDI with separated beds. Here subsequent compartments filled with either cation- or anion-exchange resins are rinsed in series by the feed water. This means that the same flow and the same current passes through the subsequent compartments. The amount of regenerating ions produced is determined by the applied electrical current, which makes the process easier to control.

The use of separated beds, electrochemically regenerated in continuous mode, was proposed in a patents of P. Kollsman [135] and others [184,194,144]. Here the water dissociation at the bipolar interfaces between an ion-exchange bed and a membrane of counter-polarity was used as the source of regenerating ions.

E. Korngold et al. [185] have proposed a concept of using two separated beds of resins, where the anion-exchange resin is fixed between two anion-exchange membranes and the cation-exchange resin between two cation-exchange membranes with a concentrate compartment placed between the two diluate compartments. H^+- and OH^-- ions produced at the electrodes migrate from electrode compartments through the membranes to the beds and regenerate the resins. A similar set-up was also used in [176,186]. Other concepts of CEDI with separated beds, proposed by H. Neumeister et al. [187], also use water splitting at electrodes, but now the electrodes are in direct contact with the corresponding ion-exchange resin beds and regenerate them directly.

E. Parsi [188,189] was among the first to propose bipolar membranes for the generation of H^+- and OH^--ions in CEDI with separated beds. In this concept the anion- and cation-exchange resin beds are separated by a bipolar membrane as shown in **Fig. 4.7**, and separated from the concentrate compartment by corresponding anion- and cation-exchange membrane. The combined use of water splitting at the electrodes and in bipolar membranes is also known [190,191].

In works of S. Thate [192,193] the CEDI with mixed-beds and two CEDI concepts with separated beds using either electrodes or bipolar membranes for the generation of H^+- and OH^--ions have been compared. The results clearly show a better removal of weakly dissociated acids by using separated beds. In this comparison the CEDI stack with bipolar membranes has performed better than the other stacks.

Compared to mixed-beds the separated bed concept has somewhat higher electrical resistance, but no reverse junctions. This allows the utilization of thicker diluate compartments filled with ion-exchange resin and reduces the number of compartments in the whole stack for a given productivity. Another important advantage of separated beds compared to mixed-bed is a better weak acids removal due to the pH changes if the anion-exchange resin compartment is situated after the cation-exchange resin compartment [194].

The three kinds of CEDI with separated beds are schematically shown in **Fig. 4.6** - **Fig. 4.8**. They differ in the source of regenerating ions which generated:

- at the bipolar *resin/membrane* interface;
- at the electrodes; or
- in the bipolar membrane.

4.4.1   Water splitting at the resin/membrane interface

CEDI with water splitting at the bipolar *resin/membrane* interface is shown in **Fig. 4.6**. The figure shows three main alternatives, where water dissociation takes place: *a*) between the cation-exchange membrane (CM) and the anion-exchange resin (AR), *b*) between cation-exchange resin (CR) and anion-exchange membrane (AM), and *c*) both between CM/AR and CR/AM.

The water dissociation takes place at the contacts between CM/AR in the **Fig. 4.6***a* and CR/AM in the **Fig. 4.6***b* and produced H^+- and OH^--ions conduct the electrical current through the respective ion-exchange materials and regenerate them. The same amount of H^+- and OH^--ions is generated in these concepts for the regeneration of CR and AR respectively. Since the intensity of water dissociation strongly depends on the catalytic activity of ion-exchange materials and on the contacting area between them, the performance of the concepts shown in the **Fig. 4.6***a* and **Fig. 4.6***b* is dependent on the nature of the membrane, the resin and the ion composition of the feed water.

The schematic drawing of **Fig. 4.6***c* shows that water dissociation occurs simultaneously at the CM/AR and CR/AM boundaries. It shows further that a concentrate compartment is located between every CR and AR filled diluate compartment and that the two concentrate compartments in the unit are not identical. In the first one, the neutralization of produced H^+- and OH^--ions takes place, while in the other one the ions removed from feed are concentrated as usual. This concept is of less practical interest for deionization, because regeneration degree of anion-exchange resin and cation-exchange resin is not always the same, the efficiency is rather low due to the partially neutralization of H^+- and OH^--ions in the concentrate compartment and because this concept requires more compartments and membranes than other concepts considered above.

Concepts similar to those in the **Fig. 4.6**, but using only one diluate compartment filled only with AR or CR are proposed for the removal of only anions or only cations [194].

4.4.2 Water splitting at the electrodes

The standard potentials $\Delta\phi$ for the water splitting due to electrode reactions are given in Eqs. (4.3) - (4.5):

Cathode: $H_2O + \bar{e} \rightarrow OH^- + \frac{1}{2} H_2$ $\qquad \Delta\phi = 0{,}42 \text{ V},$ $\qquad$ (4.3)

Anode: $\frac{1}{2} H_2O - \bar{e} \rightarrow H^+ + \frac{1}{4} O_2$ $\qquad\qquad \Delta\phi = 0{,}81 \text{ V},$ $\quad$ (4.4)

Total: $1\frac{1}{2} H_2O \rightarrow H^+ + OH^- + \frac{1}{2} H_2 + \frac{1}{4} O_2$ $\quad \Delta\phi = 1{,}23 \text{ V}.$ $\quad$ (4.5)

A CEDI scheme using the electrode reactions ("CEDI-El") is shown in **Fig. 4.7**. A unit consists of a CR-filled diluate compartment next to an anode, an AR-filled diluate compartment next to a cathode and a concentrate compartment in the middle [187]. In a stack more than one unit bipolar electrodes are used and each unit is repeated in mirror symmetry.

It is evident from Eq. (4.5) that the minimal voltage drop required to carry out the water splitting reaction at the electrodes is 1,23 V, which is higher than for direct water dissociation (Eq. (2.3)), since part of the electrical energy in the electrode reactions is used for the production of gases, H_2 and O_2.

In spite of the very good deionization performance, CEDI-El has also some disadvantages for certain applications since the gases formed in the electrode reactions are present in the diluate. Additionally, the use of corrosion-stable electrodes in each repeating unit of a stack increases the material cost for large scale applications.

140

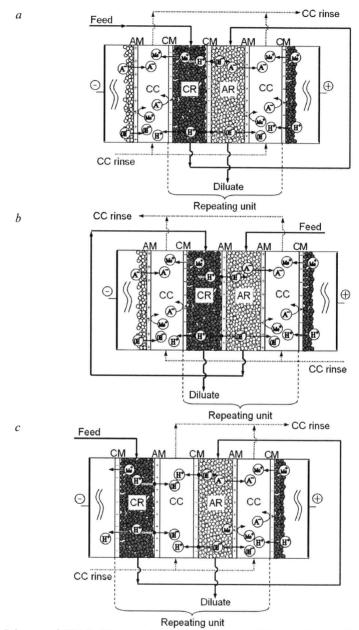

Fig. 4.6. Schemes of CEDI with separated beds with water splitting at the boundaries between ion-exchange resin and ion-exchange membrane: *a* – CM/AR; *b* – CR/AM; *c* –CM/AR and CR/AM. (CC – concentrate compartment; AM and CM – anion- and cation-exchange membrane; AR and CR – anion- and cation- exchange resin)

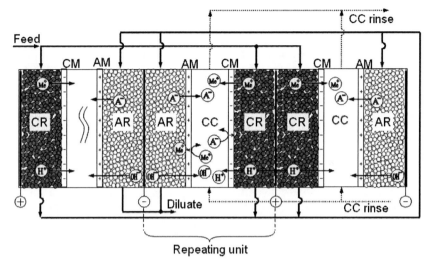

Fig. 4.7. Schematic drawing illustrating CEDI with separated beds with water splitting located at the electrodes [190]. (CC - concentrate compartment; AM and CM – anion- and cation-exchange membrane; AR and CR – anion- and cation-exchange resin)

4.4.3   CEDI with bipolar membranes

The third mentioned concept of CEDI with separated beds uses bipolar membranes to split water molecules into H^+- and OH^--ions. It is schematically shown in **Fig. 4.8**.

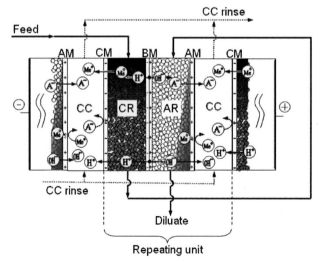

Fig. 4.8. Scheme of CEDI-BM stack according to reference [188]

As mentioned in Chapter 2.5.2, bipolar membranes have been developed to provide H^+- and OH^--ions for various electromembrane processes. A catalytic layer inside the bipolar membrane increases the water dissociation rate by a few orders of magnitude.

The CEDI with bipolar membranes ("CEDI-BM") as shown in **Fig. 4.8** has different advantages comparing to CEDI-El such as more efficient water dissociation, no electrode gases in the diluate, and the simplicity of a stack with many repeating units in one module.

4.4.4   Ion transport

Four main steps of ion transport can be distinguished in a typical CEDI with separated beds with a cation-exchange resin bed followed by an anion-exchange resin bed as shown in **Fig. 4.8**.

1. Ion-exchange in the cation-exchange resin.

In the cation-exchange bed, ion-exchange between cations (Me^+) from the feed and H^+-ions from the resin takes place:

$$\overline{CR^-H^+} + Na^+ \rightarrow \overline{CR^-Na^+} + H^+, \tag{4.6}$$

where $\overline{CR^-}$ indicates the fixed ions of the cation-exchange resin.

2. Migration of cations.

Me^+-ions, as well as not exchanged H^+-ions migrate through the cation-exchange bed and the cation-exchange membrane into the concentrate compartment. The diluate, now containing strong and weak acids, leaves the cation-exchange bed and enters the anion-exchange bed.

3. Ion-exchange in the anion-exchange resin.

The anions of strong acids (A^-) are first exchanged with OH^--ions of the anion-exchange resin:

$$\overline{AR^+OH^-} + A^- \rightarrow \overline{AR^+A^-} + OH^-, \tag{4.7}$$

where $A^- = SO_4^{2-}$; Cl^-; NO_3^-.

The ion-exchange equilibrium of this step is shifted to the right by the recombination of the OH^--ions with the H^+-ions present in the diluate (Eq. (4.2)).

In the acidic water entering the anion-exchange bed the dissociation of weak acids is rather depressed and they can be considered as non-dissociated. Such weak acids

(H*Wa*) will be absorbed by the anion-exchange resin beads in OH⁻-form, where they dissociate as follows:

$$\overline{AR^+OH^-} + HWa \rightarrow \overline{AR^+Wa^-} + H_2O,$$

where $HWa = H_2CO_3$; H_4SiO_4; H_3BO_3.

4. Migration of anions.

The anions of strong and weak acids as well as not exchanged OH⁻-ions migrate further through the anion-exchange resin and membrane into the concentrate compartment.

The ions removal in steps *1* and *3* can be described as discussed in the Chapter 2.2. Generally, the film-diffusion model will be used. The mathematical description of the transport of ions is rather complex because the changing shape of the ion profiles in the beds have to be considered not only in flow direction but also perpendicular to the flow direction. This will be discussed in more detail in Chapter 5.1. Due to the fact that the removal of weak acid anions is hindered by the low dissociation degree, the ion-exchange of Me⁺- and A⁻-ions takes place mostly in the entrance part of the cation- and anion-exchange resins, and the weak acid anions are only taken up when the majority of strong acid anions is removed.

In the CEDI with separated beds the co-ions penetrating the membrane into the first diluate compartment can be completely removed in the second diluate compartment. For a sequence of a cation-exchange bed followed by an anion-exchange bed (as shown in **Fig. 4.7** and **Fig. 4.8**), anions penetrating the cation-exchange membrane into the compartment with cation-exchange resin will be transported with all other anions of the feed into the second diluate compartment filled with anion-exchange resin where they are removed. The co-ions, i.e. cations penetrating through the anion-exchange membrane into the second diluate compartment which is filled with the anion-exchange bed will remain in the diluate. Therefore, the co-ions transport through the membranes limits the minimum diluate conductivity and this effect can be more significant in CEDI with separated beds than in the other stack concepts.

144

4.5 State of the art of commercially available CEDI

Since its initial commercialization in 1987 [150] a number of improvements concerning the design and operation of CEDI stacks have been made by different manufactures. Stacks of different capacity, covering flow rates of a few liters per hour up to 2 - 5 m³/h are commercially available today. Modifications of the original CEDI stack design have improved the efficiency of the process significantly [195,196]. Also engineering optimizations in the assembling of modules, manifolds and peripheral devices have led to compact design and coherent operation of complete systems [197,198]. Two types of CEDI module design available on the market are plate-and-frame and spiral wound modules.

The CEDI module with a typical plate-and-frame construction is composed of repeating units consisting of an anion-exchange membrane (AM), a diluate compartment, a cation-exchange membrane (CM), and a concentrate compartment between electrode compartments with electrodes and end-plates, as schematically shown in **Fig. 4.9**.

Many different improvements in the flow distribution inside and between the compartments, in sealing, as well as in the design of single parts and a whole module have been achieved by different companies in the last 20 years [199].

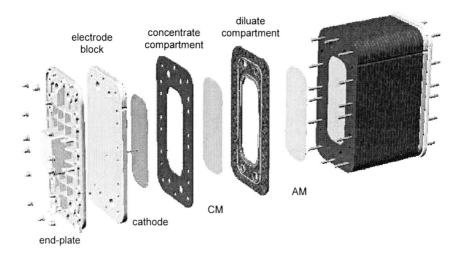

Fig. 4.9. Construction of a plate-and-frame module (exploded view of an Ionpure LX module [142])

In spiral wound CEDI modules one electrode is located in the center of the module, while the other electrode has the form of an outer cylinder. In case when the diluate flows along a spiral pathway, a very long deionization path is obtained inside a relatively compact module. In this case differences in the conductivity along the diluate path do not result in strong differences in the current density distribution but are favorably equalized [200,201]. This leads to a better deionization, especially concerning the removal of weakly dissociated compounds.

Improvements compared to the original design (**Fig. 2.22**) have been achieved by increasing the electrical conductivity of the concentrate compartment, in order to decrease the voltage drop over the stack. Initially salt was added to the concentrate rinse or the concentrate rinse was circulated in a feed-and-bleed mode [202,203]. Meanwhile concentrate compartments are often filled with ion-exchange resin [195,196,204], which significantly reduces the electrical resistance of the stack. In addition such concentrate compartments can be rinsed with low conductivity water, such as RO-permeate, which decreases the leakage of ions through the membrane from concentrate into the diluate compartment and reduces the scaling risk.

In some CEDI-stacks the electrode compartments are also filled with ion-exchange resin [205]. In other concepts the cathode compartment is filled with carbon beads [206]. A main purpose of such fillings is the minimization of the electrical resistance in electrode compartments and the prevention of scaling. The carbon beads extend the effective surface area of the cathode substantially and a more uniform distribution of the OH^--ions produced at the cathode can be achieved, which reduces the local pH close to the cathode and prevents subsequent scaling.

As described in the Chapter 2.2, the H^+-ion form of a cation-exchange resin and the OH^--ion form of an anion-exchange resin have several times higher conductivities compared to any other ionic forms of the resin. These ionic forms are prevailing in the last part of a diluate compartment, where the resins have already been largely regenerated by H^+- and OH^--ions. This can lead to a non-uniform current density distribution along the length of the ion-exchange resin bed, resulting in a loss of current efficiency and deionization performance. The use of a segmented electrode, by which the current through the different parts of the module can be adjusted, is described in [207]. The results show better ion removal when a higher current is applied to the segment at the inlet part than to the segment at the outlet part of the feed water flow path.

Other improvements of commercial CEDI stacks have been achieved by replacing the mixed-bed in the diluate compartment by partially or completely separated beds of CR and AR. The CEDI with layered beds in the diluate compartment [136,176] as illustrated in the **Fig. 4.5** was successfully commercialized and further improved by Ionpure [177,196].

CEDI modules with separated beds were introduced by „SG Wasseraufbereitung und Regenerierstation GmbH" under the name El-Ion cell, the repeating unit of which is like one shown in **Fig. 4.7**. The unit contains separated cation- and anion-exchange beds in the electrode compartments, while the generation of H^+- and OH^--ions takes place at the electrodes. The concentrate compartment, filled with an ion-exchanger is located between the diluate compartments.

Due to the separated beds of ion-exchange resins excellent removal of the weakly dissociated electrolytes, such as hydrocarbonates, silicates or borates is claimed for the El-Ion systems. The concentrate compartment filled with ion-exchange resin provides a higher electrical conductance of the stack and allows the use of RO-permeate for rinsing. The ultrapure water obtained contains H_2 and O_2, produced at the electrodes. In some cases the dissolved oxygen has a positive influence on the further storage of water in tanks, playing a role of an oxidizing agent for preventing the growth of microorganisms, however, often the presence of gases in the diluate is considered as disadvantage and may require a downstream degassing. The explosive mixture of hydrogen and oxygen creates also a safety risk and could require degassing after each electrode compartment for large scale applications.

The electrodes made from suitably durable materials are usually expensive and add significantly to the material costs of a stack. This is one of the reasons why CEDI-El is mostly used in small scale units with modules having no more than two or three electrodes. For an assembly of more repeating units and for the production of gas-free diluate the use of CEDI with separated beds with bipolar membranes, like illustrated in the **Fig. 4.8** is more reasonable.

A CEDI-El module with two repeating units in one stack had been manufactured under the name ADEPT (Advanced Deionization by Electrical Purification Technology) by Elga Ltd., which uses a sequence of separated beds in series with generation of H^+-ions at one anode and OH^--ions at two cathodes as shown in **Fig. 4.10**. The advantage of this design is the deionization in two subsequent stacks. Most of the ions are removed in the first stack (**Fig. 4.10** below), yielding a concentrate with high ion concentration. In the second stack (**Fig. 4.10** above) the final purification takes place in contact with two concentrate compartments, which are rinsed with RO-permeate. This prevents high concentration gradients over the membrane between concentrate and diluate compartments and subsequent contamination of diluate with co-ions in the final polishing stack. The dividing into two stages also allows to separately control the current and/or voltage applied on both stacks and leads to a more even current density distribution inside a stack. The disadvantage is again the high concentration of electrode gases in the product and a more complicated construction.

CEDI with separated beds and bipolar membranes has been studied extensively in experiments and mathematical simulation in the PhD thesis of Sven Thate [193] and was published in several articles [192, 208-210]. Some conclusions of this work will be reconsidered in Chapter 5.

148

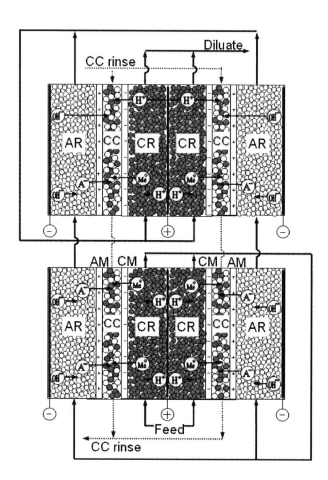

Fig. 4.10. Two stacks comprising of two unit cells of CEDI with separated beds of anion-exchange (AR) and cation-exchange (CR) resin using electrode reactions for resin regeneration, where concentrate compartments (CC) between anion-exchange membrane (AM) and cation-exchange membrane (CM) are filled with an ion-exchanger and RO-permeate is used to feed diluate and concentrate compartments (prospect information: Water technology for process applications. Elga Labwater, 1996)

4.6   Summary of literature study about CEDI

Since the first suggested concept of CEDI with a mixed bed in the diluate and a neutral screen in concentrate compartment, substantial modifications and improvements have been published. Currently, the filling of concentrate compartments with an ion-exchange resin is used by most manufacturers of CEDI modules. This has improved the performance of CEDI by providing better purification and stability to scaling, as well as reduced energy consumption. Also the filling of electrode compartments by an ion-conductive or electron-conductive packed bed leads to similar improvements, especially reducing the risk of scaling, which is more effective when an anion-exchange material is filled in the anode compartment and an electron-conductive material like carbon beads in the cathode compartment.

The further introduced concepts with clustered or layered beds of ion-exchange resins in the diluate compartment, or of cation- and anion-exchange beds in separated compartments prevent the formation of reverse junctions between resin beads. This allows a more efficient removal of weakly dissociated electrolytes. The absence of reverse junctions and the higher conductivity of the beds allow using compartments which are thicker than in case of mixed beds. This reduces the investment costs and the pressure drop over the beds.

The present trend in the development of CEDI concepts goes to separated beds of cation- and anion-exchange resins in the diluate compartment, which provides excellent removal of strongly as well as weakly dissociated electrolytes. Moreover, passing feed water first through the cation-exchange resin bed leads to an acidification of the water which prevents fouling and scaling formation.

The main milestones of CEDI development over the past twenty years are summarized in the **Table 4.1**.

As a result of continuous improvements, CEDI can produce water with high resistivity, while investment costs for CEDI modules were significantly reduced and CEDI became competitive in applications where mixed bed deionization was previously used. The operating costs for CEDI are significantly lower than that for mixed-bed deionization, mainly because no regeneration chemicals are needed and less maintenance is required.

Table 4.1

Modifications in CEDI since the first concept

Modification	Achievement
Rinsing of the concentrate compartments with RO-retentate or addition of salt	Increase of stack conductivity
Filling of the concentrate compartments with mixed-bed	Increase of stack conductance; Decrease of scaling and fouling risk; Reverse of polarity possible
Filling of cathode compartments with carbon beads or ion-exchange resin	Increase of stack conductance; Decrease scaling risk
Use of clustered or layered beds	Better removal of strongly and weakly dissociated compounds; Increase of stack conductance; Allows the use of thicker compartments
Use of separated beds with generation of H^+- and OH^--ions at the electrodes or in bipolar membranes	Better removal of strongly and weakly dissociated compounds; Increase of stack conductance

Nonetheless, the increasing demand on water quality and quantity and the need to reduce the energy and water consumption as well as other cost factors, create needs for further improvements of the CEDI technology. The main requirements for improvement can be listed as follow:

- better product water quality with removal of silica and boron to sub-ppb level by CEDI without downstream polishing;
- reliable operation under conditions of feed water with a certain hardness, higher conductivity, and/or high CO_2 content. Accepting strong variations of feed water quality in time. Short start-up time to deliver the required water quality;
- lower investment costs;
- lower operating costs by reduction of energy and water consumption;
- meeting the norms and guidelines for specific applications, like possibility of in-line sterilization for pharmaceutical production and medical applications.

Further improvements also require optimization of the whole value chain of water purification. For the production of ultrapure water in particular the coupling of the pretreatment by reverse osmosis (RO) and subsequent CEDI will require adjustment of the flow rates and currents in the CEDI module to the flow rate, recovery rate and efficiency of the RO [211]. Also combinations of CEDI with other techniques and their optimization could change performance and costs of the complete system. The technologies, which can be used upstream CEDI are degassing in order to obtain lower CO_2 levels in CEDI feed [115], nanofiltration instead of RO in order to reduce power consumption and oxidation techniques, like UV, for better removal of organic matters.

Concerning CEDI itself, the choice and the further optimization of the presently most effective concepts can help to further improve its performance. Based on the work of S. Thate [193] the CEDI concept with separated beds with water dissociation in bipolar membrane has been selected for this task, which is the topic of the next chapter.

5 Improved concepts for CEDI with bipolar membranes

The review of existing CEDI technologies in Chapter 4 has shown that among the available CEDI concepts, CEDI with separated beds and bipolar membranes is best suited to produce ultra pure water with a minimum amount of weakly dissociated compounds. This has already been the reason for a detailed investigation of CEDI-BM systems by S. Thate [193], who combined his experiments with detailed mathematical modeling and simulation. The following work is to a large extent based upon his studies. Their main findings will therefore be briefly reviewed in the following section.

5.1 Previous work and present challenges

In his work S. Thate compared CEDI-BM (see **Fig. 4.8**) with different flow sequences of cation and anion exchange resin beds in series. He has shown that the flow first through the cation-exchange resin (CR) bed and second through the anion-exchange resin (AR) bed (CR→AR sequence) leads to a better deionization, especially regarding weak acids, than the opposite sequence. S. Thate has based the preference of the CR→AR sequence on several reasons, listed in the following.

1. In the first bed cations are taken up and H^+-ions are released, leading to an acidic feed to the second bed. The anion-exchange equilibrium following Eq. (4.7) is therefore shifted to the right in the second bed, since the OH^--ions released recombine with the available H^+-ions to water, as explained in Chapter 4.2.

2. The pH shift is particularly advantageous for the removal of weakly dissociated acids. Due to the low pH these compounds leave the first bed undissociated. Due to the high pH inside the anion-exchange beads which are in OH^--form in the second bed, even previously non-dissociated acids dissociate and can be removed.

3. Downstream the cation-exchange resin bed, the lower pH of water entering the anion-exchange resin bed results in a higher fraction of HCO_3^--ions inside the anion-exchange beads compare to the AR→CR sequence, where a higher fraction of CO_3^{2-}-ions is expected. Since the anion-exchange resin has only half the capacity for the uptake of the bivalent form than for the univalent and since the univalent HCO_3^--ions

have a higher mobility inside the anion-exchange resin, an easier transport of HCO_3^--ions through the anion-exchange bed and more advantageous current density distribution by the CR→AR sequence occur.

4. Due to the acidification of water in the first bed, this sequence also has a higher stability against scaling, even if feed water containing hardness ions is used, whereas the opposite sequence would cause hardness ions to precipitate at the high pH of the first bed.

5. The above explanation could also be linked to the experimental observation, that the sequence CR→AR is more stable against organic fouling than the AR→CR sequence, if a feed water containing humic acids is used.

As a shortcoming of his experiments, S. Thate observed a certain leakage of cations from the concentrate through the anion-exchange membrane into the diluate in the anion-exchange bed (see **Fig. 4.7** and **Fig. 4.8**), due to limited anion-exchange membrane selectivity. In most of his experiments S. Thate used concentrate compartments with conventional inert spacers, rinsed by softened tap-water or RO-permeate. Some short term tests comparing concentrate compartments filled with either an anion- or a cation-exchange resin showed that the concentrate compartment filled with anion-exchange resin clearly gave the lowest diluate conductivity.

These results have been the starting point of the following investigations. Their goal is to show the influence of different configurations of CEDI-BM on the conductivity of the diluate produced at different operating conditions and for different concentrations of feed and concentrate. Based on the results, new design concepts for an improvement of the process are proposed.

Three different concepts of CEDI-BM will be experimentally compared in the following. The first concept, where the concentrate compartment of a CEDI-BM is filled by a neutral spacer, is presented in **Fig. 4.8**. This is the standard design described by E. Parsi [188] and studied by S. Thate [193].

The second concept of CEDI-BM has an additional compartment, called protection compartment (PC) which separates the anion-exchange membrane of the diluate compartment from the concentrate compartment. As shown in the **Fig. 5.1** [212], the protection compartment is filled with an anion-exchange resin and is separated from the concentrate compartment by a second anion-exchange membrane (^{PC}AM). Filling the protection compartment with ion-exchange resin allows using low conductivity water, for example a part of the diluate produced, to rinse this compartment, without significantly increasing the electrical resistance of the stack. A low conductivity of the rinse solution is necessary to prevent any co-ion leakage into the diluate compartment. Co-ions (Na^+) penetrating the ^{PC}AM from the concentrate compartment will be present in water at the right side along the flow path of the protection compartment as indicated schematically by the grey area in the **Fig. 5.1**. This distribution is a result of the interplay of the upwards flowing PC-rinse, which tends to sweep the co-ions out and their electromigration (and diffusion) through the ^{PC}AM from right to left. The goal is to rinse the protection compartment with diluate at such a rate, that a profile of co-ions as depicted by the grey area in **Fig. 5.1** is established, without any co-ions leakage into the diluate compartment.

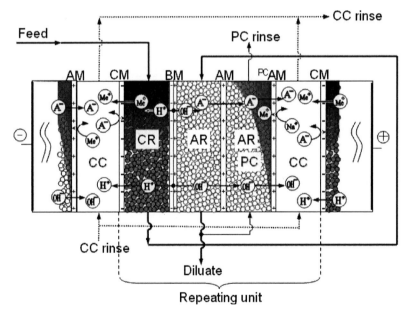

Fig. 5.1. General concept of CEDI-BM with a protection compartment (PC) filled with anion-exchange resin (AR)

The third concept used in the experimental investigation has some similarity with the stack concept already briefly studied by S. Thate. Here the concentrate compartment (CC) is filled with an anion-exchange resin, as shown in the **Fig. 5.2**.

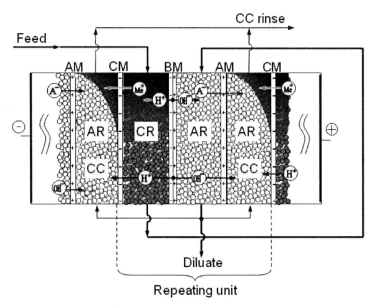

Fig. 5.2. Scheme of CEDI-BM with the concentrate compartment filled with anion-exchange resin (AR)

In the scheme of the **Fig. 5.2** the anion-exchange resin in the concentrate compartment plays a similar role as in the protection compartment described before. Different from the preliminary tests of S. Thate, the concentrate compartment is now rinsed with a small part of the produced diluate. Compared to the CEDI-BM-Pro unit, the concentration of Na^+-ions will now rise to substantially higher concentrations than in the protection compartment, since cations migrate freely through the cation-exchange membrane (CM) into the concentrate compartment. This means that the rinse should be higher to avoid high Na^+-concentrations close to the anion-exchange membrane (AM), separating the concentrate compartment (CC) from the second diluate bed with anion-exchange resin (AR). A desired distribution of cations is again marked by the grey area.

The advantage of the concept depicted in **Fig. 5.2** over that shown in **Fig. 5.1**, is that only three compartments (two diluate and one concentrate) are combined into one unit cell as compared to four compartments (two diluate, one protection and one concentrate) in the process illustrated in **Fig. 5.1**.

5.2 Experimental set-up

5.2.1   Materials

The anion-exchange, the cation-exchange and the bipolar membranes used in the experiments are the Neosepta membranes of the types AMH, CMB and BP-1, produced by Tokuyama Corporation. These membranes are manufactured by the "paste method" [44,45,47] and have a microheterogeneous structure, which means a higher permselectivity compared to heterogeneous membranes.

The ion-exchange resins used in the tests are the Dowex Monosphere A550 UPW and the Dowex Monosphere C650 UPW, respectively. They are resins of ultrapure water production grade, manufactured by Dow Chemical Company.

The experimental investigation of CEDI was carried out in the same plate-and-frame stacks made out of Plexiglas as used by S. Thate [193]. The electrodes consist of a platinum coated titanium screen.

The tubing used for the feed and for the concentrate stream is made of PTFE and PVC, respectively. A polyurethane tubing of special purity grade (290 PUR-Ether Tubing, Nalgene®) was utilized for the diluate stream in order to prevent possible desorption of impurities from the tubing into the water as well as diffusion of CO_2 from the ambient air into the water.

5.2.2   Stacks

The assembled stacks have an effective membrane area of 100×300 mm (W×H). The thickness of each compartment is 10 mm. The sealing between the frames and the membranes is achieved by rubber O-rings. In some experiments the electrodes of one side of the stack are divided into ten parts, which were separately connected to the same voltage sources as shown on the photographs of the **Fig. 5.3**. This allows measuring the current density distribution along the flow path of the stack.

a   b

Fig. 5.3. Photographs of the experimental cell: *a* – end-plate with an electrode segmented in ten parts; *b* – assembled stack with ten outside connections of the electrode

The distance between an electrode and its neighbor membrane was adjusted by conventional spacers of 0,6 mm thickness. Due to this short thickness the resistance of the electrode compartment is more than one order of magnitude lower, compared to the resistance of the stack. This made it possible to draw conclusions for the current density distribution in the stack from the measured current density distribution at the electrodes, although some deviation by non-parallel current flow in the electrode compartment and in the stack cannot be excluded. Such deviations can result from strong differences in resistivity along the flow path of the stack.

Three different CEDI stacks with bipolar membrane were assembled as schematically illustrated in **Fig. 5.4** and **Fig. 5.5**. Each stack in **Fig. 5.4a** and **Fig. 5.4b** consists of one deionization unit composed of two diluate compartments placed on both sides of the bipolar membrane (BM). The diluate compartments are filled with cation-exchange resin (CR) and confined by a cation-exchange membrane (CM) from the cathode side and filled with anion-exchange resin (AR) and confined by an anion-exchange membrane (AM) from the anode side. In these stacks the electrode compartments also play the role of concentrate compartments. The stack assembled corresponding to **Fig. 5.4a** represents the concept suggested by E. Parsi [188], which has also experimentally been investigated in the work of S. Thate [192,193]. It will further be referred to as CEDI-BM.

The scheme in **Fig. 5.4**b differs from the CEDI-BM by an additional protection compartment (PC) placed between the AM of the AR-diluate compartment and the concentrate compartment (CC) at the anode. The protection compartment is filled with AR and separated from the concentrate compartment by an AM. This CEDI-BM with protection compartment is further referred to as CEDI-BM-Pro. Both CEDI-BM and CEDI-BM-Pro stacks were experimentally tested separately as well as simultaneously with identical feed water to investigate the influence of different process parameters.

For a further series of experiments another stack was assembled in order to compare a unit cell with a protection compartment (CEDI-BM-Pro) with a unit cell where the concentrate compartment was filled with anion-exchange resin. This third stack is schematically illustrated in **Fig. 5.5**. A part of the diluate produced in the respective unit cell was used to rinse the PC or the CC with the flow rate of 2,5 L/h each.

The feed water streams are directed from top to bottom first through the CR bed and then through the AR bed of the diluate compartments. The electrode rinse flows from bottom to top first through the anode compartment and then through the cathode compartment.

160

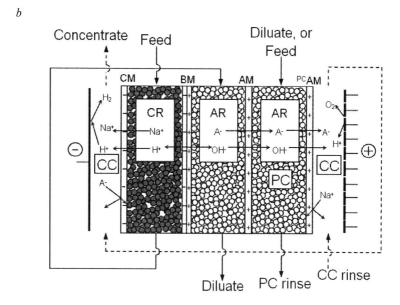

Fig. 5.4. Scheme of two investigated stacks: a – CEDI-BM; b – CEDI-BM-Pro. Here AR and CR relates to anion- and cation-exchange resin; AM, CM and BM is anion-, cation- and bipolar membrane; CC and PC is concentrate and protection compartment respectively

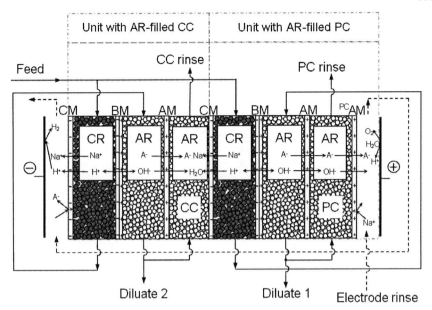

Fig. 5.5. Scheme of the investigated stack consisting of two units, one with a protection compartment (PC) and another with the concentrate compartment (CC) filled with anion-exchange resin (AR)

5.2.3   Experimental set-up and conditions

The experimental set-up described in Chapter 3.3.3 (**Fig. 3.19**) was extended for the simultaneous tests of two modules as shown in **Fig. 5.6**.

The same water pretreatment block as shown in **Fig. 3.19** was used to feed the CEDI-BM modules. Stuttgart tap-water was softened by Na^+-ion exchange, resulting in softened tap-water with conductivity of 332 ± 3 µS/cm and pH of 7,76; After softening it contained different anions and only Na^+-ion as mineral cation (**Table 5.1**). The softened tap-water (STW) was filled into Tank 2 and used to rinse the electrode compartments of both stacks. The same softened tap-water was also used to feed the RO-unit. The RO-permeate with conductivity of 2,3 µS/cm was collected in the Tank 1 and used as feed for the CEDI stacks in most experiments, except for the studies with varying feed conductivity. In these experiments the conductivity of the RO-permeate was varied by changing the RO recovery rate through regulation of the retentate flow rate. The concentration of Na^+-ions and different anions in the softened tap-water and in the RO-permeate are summarized in the **Table 5.1**.

162

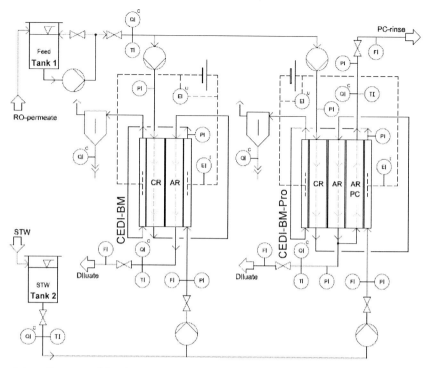

Fig. 5.6.   Set-up used for the experimental investigation of the CEDI stacks

Table 5.1

Ion-content of utilized softened tap-water and RO-permeate [a]

Concentration of ions / Water type	$C_{Na^+}\left[\dfrac{\mu eq}{L}\right]$	$C_{SO_4^{2-}}\left[\dfrac{\mu eq}{L}\right]$	$C_{Cl^-}\left[\dfrac{\mu eq}{L}\right]$	$C_{NO_3^-}\left[\dfrac{\mu eq}{L}\right]$	$C_{Si}\left[\dfrac{\mu mol}{L}\right]$	pH	κ, μS/cm
Concentrate (softened tap water)	3500	600	170	64	not measured	7.8	332
Feed (RO-permeate)	20	0.38	0.86	2.24	4.8	5.8	2.3

[a] The difference between the sodium concentration and the sum of analysed anions in the RO-permeate is the HCO_3^- concentration.

If not additionally specified, the flow rate of softened tap-water, which rinses the anode and cathode compartments in series was set to 15 L/h and the feed flow rate in the stacks was $Q_F = 48$ L/h. The pressure drop over the diluate path in both, the cation- and anion-exchange resin, compartments at 48 L/h was ca. 1 bar. In the experiments either a part of the diluate or the RO-permeate was utilized to rinse the PC. In the experiments with the CEDI-BM-Pro and in the comparative experiments of CEDI-BM and CEDI-BM-Pro the hydrostatic pressure drop between the outlet of anion-exchange resin filled diluate compartment in CEDI-BM (or PC in CEDI-BM-Pro) and the inlet of anode compartment was kept to be ca. 50 mbar. Since the pressure in the anode compartment was close to ambient, the higher pressure in the adjacent compartment filled with anion-exchange resin prevented a convective transport of contaminants from the concentrate through the anion-exchange membrane.

The experiments were carried out at constant electrical current provided by the direct current power supply ISO-Tech IPS 1603D. The voltage drop over the stack and the currents through each separated section of the electrodes were recorded continuously. The flow rates and the pressures in the different compartments were periodically controlled.

The temperatures of feed and concentrate were measured, but not controlled. The feed water had a temperature of about 16°C in winter and about 20°C in summer.

The conductivities referenced to 25°C were measured online in the feed, the diluate, the concentrate and the PC-rinse streams. Although the conductivity measurements do not give quantitative information about the content of individual ions or non-dissociated impurities in the diluate, they are a convenient and quite satisfactory to compare the deionization performance of the different CEDI stacks.

Experimental conditions deviating from the ones defined above will be specified in the corresponding experimental part. The performed series of experiments are summarized in the **Table 5.2**.

Table 5.2

Summary of carried out experiments

Stack	Varied parameters	Temperature, °C
CEDI-BM	pressure in CC flow rate in CC	16°C
CEDI-BM vs. CEDI-BM-Pro (comparative)	current	16-18°C
CEDI-BM-Pro	flow rate in diluate flow rate in PC	16-18°C
	rinse-water of PC flow directions	21-22°C
Two-units stack	current feed conductivity	18-22°C

It is a general experience that the establishment of a steady state in CEDI experiments can take quite long. This is due to the relatively low ion concentration in the RO-permeate, and the high capacity of ion-exchange resins in the stacks. In the experiments presented the time to reach a steady-state required generally between 3 to 15 days of continuous operation. The steady-state was confirmed, when the stack voltage, the current density distribution at the electrodes, and the conductivities in the outlet of the diluate, the concentrate and the protection compartments have been constant for 24 hours.

5.2.4   Conductivity measurements

The conductivity measurements were performed with a microprocessor conductivity meter LF3000 combined with an extension unit Multiplex3000/LF (WTW, Germany) allowing simultaneous measurements with seven conductivity cells. The utilized conductivity cells are: LR 01/T in the feed, LR 001/T in the diluate and KDU 1/T in the concentrate streams. The conductivity meters automatically perform a non-linear temperature compensation of the specific conductivity to the reference temperature of 25°C.

5.3 Results and discussion

The results of the deionization experiments with the three different stacks, performed with varying process parameters are presented and discussed below. Simplified balance equations for the co-ion transport through ion-exchange membranes are applied for the interpretation of some results obtained experimentally.

5.3.1   Influence of pressure and flow rate in the concentrate compartment on the diluate conductivity in the CEDI-BM

To estimate the influence of pressure and flow rate in the concentrate compartment on the diluate conductivity two sets of experiments were carried out with the CEDI-BM, stack shown in **Fig. 5.4a** at a current density of 33 A/m^2 and a feed conductivity of 2,6 μS/cm.

The flow rate of the diluate was 48 L/h resulting in a pressure drop in the AR-filled diluate compartment from about 0,048 MPa at its inlet to about zero at the outlet. The Dowex Monosphere C500 was used as cation-exchange resin, which is very similar to Dowex Monosphere C650 utilized in all following experimental tests.

In one set of experiments the flow rate in the concentrate compartment was kept constant and the pressure in the concentrate compartment (P_{CC}) has been regulated by the pump pressure and the valve at the outlet of the concentrate compartment. In another set of experiments the flow rate in the concentrate compartment and the pressure in the concentrate compartment are changed simultaneously, as shown in **Fig. 5.7**.

It can be noted from the small and almost constant difference between the inlet (P_{CC}^{IN}) and the outlet pressure (P_{CC}^{OUT}) that the concentrate compartment, containing a net-electrode and a spacer of open mesh type, has a very low hydrodynamic resistance. Thus, the increase of the average pressure in the concentrate compartment with flow rate is mainly due to the pressure losses in the outlet manifold of the concentrate compartment. The dependence of the average pressure P_{CC} [MPa] upon the flow rate Q_{CC} [L/h] is shown as dashed line in **Fig. 5.7**. It can be approximated by:

$$P_{CC} = 2{,}07 \cdot 10^{-5} Q_{CC}^2 + 1{,}04 \cdot 10^{-4} Q_{CC} + 5{,}4 \cdot 10^{-3}. \tag{5.1}$$

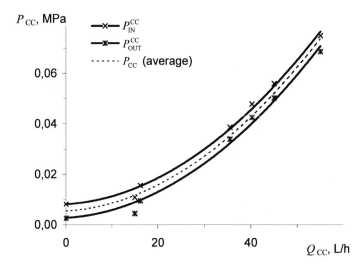

Fig. 5.7. Influence of the flow rate Q_{CC} on the pressures P_{CC}, measured in the concentrate compartment

166

Fig. 5.8 shows the dependence of the diluate conductivity upon the average pressure in the concentrate compartment. The results of both experimental sets in **Fig. 5.8** indicate the sensitivity of the diluate conductivity upon the pressure and the flow velocity in the concentrate compartment.

The increase of the pressure in the concentrate compartment from ca. 0,005 MPa to 0,08 MPa at constant flow rate leads to a significant increase in the diluate conductivity from 0,17 µS/cm to 0,3 µS/cm.

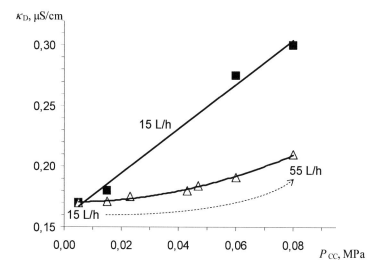

Fig. 5.8. Influence of pressure and flow rate in concentrate compartment on the diluate conductivity in CEDI-BM: —■— - increase of pressure at constant flow rate in concentrate compartment (Q_{CC} = 15 L/h); —△— - increase of pressure is conjugated with the increase of the flow rate in concentrate compartment from 15 L/h to 55 L/h

As shown in [213] and [214], the pressure difference between the concentrate and the diluate compartment can increase the co-ions transport through an ion-exchange membrane. It is also shown in [215] and [216], that the co-ions transport through the membrane increases linearly with increasing pressure difference between concentrate and diluate in a range of positive ΔP, which is in accord with the nearly linear κ_D - P_{CC} dependence at a constant flow of 15 L/h (**Fig. 5.8**) and is also predicted by Eq. (2.18). The correlation between the flux of sodium as the sole co-ion and the pressure gradient over the membrane can be rewritten from Eqs. (2.11) and (2.17) as:

$$-\boldsymbol{J}_{Na^+} = \left(\underbrace{\overline{C}_{Na^+} \frac{\overline{V}_{m_{Na^+}} \overline{D}_{Na^+}}{RT}}_{pressure\ diffusion} + \underbrace{\overline{C}_{Na^+} \overline{L}_h}_{convective\ transport} \right) \nabla P . \qquad (5.2)$$

At the increasing pressure in the concentrate compartment, the flux of co-ions into the diluate, caused by the pressure diffusion and coupled with convective flux, will be increased, and the diluate conductivity will be increased also, reflecting the linear κ_D - P_{CC} dependence at the constant flow rate (**Fig. 5.8**).

It is difficult to separate the influence of the pressure diffusion and the coupled convective transport of co-ions. It can be assumed however that the pressure diffusion plays a minor role in the transmembrane transport and co-ions will penetrate through the membrane mostly with convective volume flux through the inhomogeneous regions and defects in the membrane.

Comparing both series of experiments shown in **Fig. 5.8** it is evident, that at the same pressure in the concentrate compartment the diluate conductivity is lower at higher flow rate in the concentrate. The reason is the reducing influence of concentration polarization with increasing flow rate, as explained in Chapter 2.4.2. The decrease of the co-ions boundary concentration with increasing flow rate and the resulting decrease of the concentration gradient over the anion-exchange membrane diminishes the number of co-ions penetrating the anion-exchange membrane and being transported into the diluate.

In the experiments with changing flow rate, flow velocity and pressure are coupled. This means that concentration polarization will decrease with increasing flow velocity, but the pressure and consequently the pressure driven co-ions leakage will increase. The result of this interplay is the experimentally observed curve. The form of the curve is related to the pressure driven co-ions transport, the Donnan equilibrium and the concentration polarization in the concentrate compartment, as will be shown in the following.

Simplifying the divergence of the pressure in the Eq. (5.2), the flux of sodium ions is therefore proportional to the sodium concentration in the membrane, $\overline{C}_{Na^+}$ and the pressure in the concentrate compartment, P_{CC}:

$$\left| \overline{\boldsymbol{J}}_{Na^+} \right| \propto \overline{C}_{Na^+} P_{CC} . \qquad (5.3)$$

Donnan equilibrium for a 1:1 electrolyte, results in a direct proportionality between activities (approximated by concentrations) of Na^+-cations inside the anion-exchange membrane and in the solution at the membrane surface facing concentrate compartment, i.e. $\overline{C}_{Na^+} \propto {}^{CC}C_{Na^+}^S$. The concentration of the solution at the membrane surface (${}^{CC}C_{Na^+}^S$) exceeds the bulk concentration (${}^{CC}C_{Na^+}^B$) and can be obtained from Eq. (2.41), leading to:

$$ {}^{CC}C_{Na^+}^S = {}^{CC}C_{Na^+}^B + \frac{i\delta\left(t_{Na^+} - \bar{t}_{Na^+}\right)}{FD_{El}}, \tag{5.4} $$

where D_{El} is the diffusion coefficient of the electrolyte in the boundary layer; i is the current density; F is the Faraday constant; t_i and $\bar{t}_i$ the transport numbers of an ion in the solution and in the membrane, respectively.

The average thickness δ^{av} of the boundary layer in a membrane compartment with a spacer can be approximated by:

$$ \delta^{av} = a \cdot \left(\frac{D_{El} \cdot l_D \cdot d_h}{v}\right)^n, \tag{5.5} $$

where v is the average flow velocity of the concentrate; d_h is the hydraulic diameter of the channel, which is defined as four volumes per wetted area [217]. The coefficients l_D, a, and n depend on the spacer geometry and have to be determined experimentally.

Taking into account $v \propto Q_{CC}$, and the relations specified in Eqs. (5.3), (5.4) and (5.5) the dependence of the co-ions flux upon flow rate and pressure can be described by the following general dependence:

$$ \left|\overline{J}_{Na^+}\right| = \Theta P_{CC} Q_{CC}^{-n} + \Pi P_{CC} + \Phi, \tag{5.6} $$

where Π and Θ are proportionality coefficients. Φ is the co-ion flux at the zero pressure in the concentrate compartment due to the applied electrical potential. A frequently found value for the constant n is $n = \frac{1}{3}$.

Sodium co-ions penetrating the anion-exchange membrane appear in the diluate in a concentration ${}^D C_{Na^+}$:

$$ {}^D C_{Na^+} = \frac{\left|\overline{J}_{Na^+}\right| \cdot A}{Q_F}, \tag{5.7} $$

where $\overline{J}_{Na^+}$ is the flux of sodium ions through the membrane of area A, and Q_F is the flow rate of feed water through the diluate compartment.

The main anion leaving the AR diluate compartment together with the Na^+-ions is the OH^--ion, because in the outlet part of the bed the anion-exchange resin should be fully regenerated to the OH^--form. The conductivity of the diluate containing the NaOH (κ_{NaOH}) can be calculated from Eq. (5.8):

$$\kappa_{NaOH} = \lambda_{Na^+} C_{Na^+} + \lambda_{OH^-} C_{OH^-} + \lambda_{H^+} C_{H^+}, \tag{5.8}$$

where λ and C is the equivalent conductivity and the concentration of the respective ions.

C_{OH^-} and C_{H^+} can be found from the electroneutrality conditions (Eq. (5.9)), and the water dissociation constant K_W (Eq. (5.10)):

$$C_{Na^+} + C_{H^+} = C_{OH^-}, \tag{5.9}$$

$$K_W \approx C_{H^+} \cdot C_{OH^-} \approx 10^{-14}. \tag{5.10}$$

Combining Eq. (5.6) with the experimentally obtained P_{CC} versus Q_{CC} dependence, Eq. (5.1), and with Eqs (5.11)-(5.10) a relation for κ_D versus Q_{CC} is obtained. Applying the best fit for Π and Θ, this dependency is in good agreement with the results of the experiments as shown in **Fig. 5.9**.

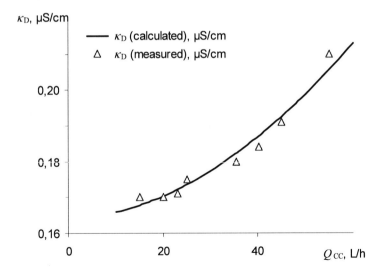

Fig. 5.9. Influence of the flow rate in the concentrate compartment on the diluate conductivity of the CEDI-BM stack, if the flow rate and the pressure change simultaneously. $\triangle$ – experimental results; line – calculated using following fitting parameters in Eq. (5.6): $\Pi = 3,15 \cdot 10^{-3}$; $\Theta = 3,95 \cdot 10^{-3}$ and $\Phi = 9,71 \cdot 10^{-4}$

The example shows that the pressure driven co-ion transport from the concentrate into the diluate plays an increasing role with decreasing diluate concentration and must be taken into account in high purity water production. To optimize the performance of CEDI-BM, the pressure in the concentrate compartment should be lower than the pressure in the next diluate compartment. This means that the hydrodynamic resistance of the concentrate flow through the compartments and downstream should be minimized. But due to their limited mechanical strength, the pressure difference over the membrane should not be too high, in order to prevent mechanical failure of the membrane.

Therefore, as mentioned in Chapter 5.2.3, the pressure in the anion-exchange resin filled diluate or protection compartment next to the anode was always kept somewhat above the pressure of the anode compartment.

5.3.2 Comparison of CEDI-BM and CEDI-BM-Pro at different current densities

The experimental series comparing CEDI-BM and CEDI-BM-Pro were carried out simultaneously with the same feed in the two stacks, schematically shown in **Fig. 5.4**. In the experiments with CEDI-BM-Pro, from 48 L/h of the diluate a part of $Q_{PC} = 10$ L/h was utilized to rinse the PC. **Fig. 5.10** shows the resulting conductivity (κ_D) and the resistivity (ρ_D) of the diluate measured at the different current densities.

It is evident from **Fig. 5.10**, that in the whole range of tested current densities the stack with the protection compartment allows to obtain a diluate with appreciably lower specific conductivity (κ_D), than the CEDI-BM stack without PC. Also the dependencies of conductivity over current density are quite different. In CEDI-BM-Pro, κ_D steadily decreases with increasing current density down to $0,0555$ μS/cm at $i = 100$ A/m^2. In CEDI-BM however κ_D first decreases and then increases with increasing current density, showing an optimum current density of ca. 27 A/m^2 for which the lowest diluate conductivity is achieved.

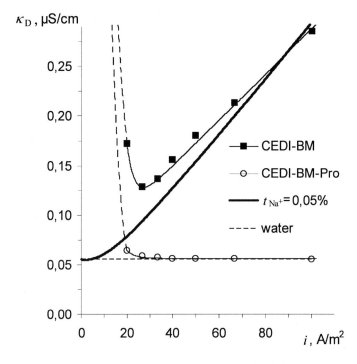

Fig. 5.10. Effect of the current density on the specific conductivity of the diluate. The probable extension of the curves to lower values of i are indicated by the dashed lines

With increasing current density, a larger amount of H^+- and OH^--ions is produced in the bipolar membrane, resulting in a higher regeneration of cation- and anion-exchange resin. This enhances the ion-exchange removal of ions from the feed water and results in the strong initial decrease of the diluate conductivity from the feed value of 2,3 µS/cm in both stacks. In the CEDI-BM-Pro stack the decrease of the diluate conductivity from the feed value to 0,065 µS/cm at a current density of 20 A/m² is due to the removal of all strongly and most weakly dissociated electrolytes. The subsequent gradual decrease of the diluate conductivity at higher current densities is due to the removal of the residual weakly dissociated electrolytes.

If identical current densities and feed flow rates in the CEDI-BM and the CEDI-BM-Pro stack are assumed, the ion distribution profiles in the diluate compartments and the amount of counter-ions removed in the diluate beds should be about identical in both stacks. Thus, it can be concluded that the subsequent increase of the diluate conductivity in the CEDI-BM stack is caused by the increased transport of co-ions, here Na$^+$-ions, through the anion-exchange membrane from the concentrate directly into the diluate, whereas the protection compartment prevents this contamination in the CEDI-BM-Pro stack.

In the present experiments the anion-exchange resin in the diluate compartment has a pure gel-structure with an ion-exchange capacity higher than that of the anion-exchange membrane. It is known that, because of its higher ion-exchange capacity, a gel-type strongly basic (or strongly acidic) ion-exchange resin has a higher Donnan exclusion of co-ions than a conventional ion-exchange membrane. This means that Na$^+$ co-ions, penetrating the membrane into the anion-exchange bed remain excluded from the resin phase and are rinsed out together with charge compensating OH$^-$-ions by the diluate stream.

The concentration of co-ions ($^D C_{Na^+}$) in the diluate of the CEDI-BM expressed by Eq. (5.7) is related to the co-ions flux, which can be simplified in this case as the flux caused by electromigration only as given by Eq.(2.43). It is proportional to the current density:

$$^D C_{Na^+} = \frac{i \cdot A \cdot \overline{t_{Na^+}^{AM}}}{F \cdot Q_F},$$

(5.11)

where i is the current density; A - the membrane area; $\overline{t_{Na^+}^{AM}}$ - the integral transport number of Na$^+$-ion across the AM; Q_F – the feed flow rate; and F – Faraday constant.

The charge of the Na$^+$-ions present in the diluate will be compensated by OH$^-$-ions, released from the regenerated anion-exchange resin. Now the conductivity of the diluate outlet in the CEDI-BM stack can be calculated by Eqs. (5.8), (5.9), (5.10) and (5.11) since it is caused only by co-ions transport. The result of such a calculation is shown in **Fig. 5.10** as a continuous line for an average Na$^+$-ion transport number $\overline{t_{Na^+}^{AM}}$ of 0,05%.

The calculated curve reflects the nearly linear increase of the diluate conductivity in the CEDI-BM stack at higher current densities reasonably well, but it is, of course, only a rough estimate of the underlying complex behavior.

Many interdependent effects influence the $\kappa_D - i$ dependence at higher current densities. An increasing current density leads to an increase in the degree of ion-exchange resins regeneration, which means a higher H^+-ions content in the CR, and a higher OH^--ions content in the AR. This results in shifting of the ionic profiles in the ion-exchange resins to the inlet of the respective compartment. Due to the higher conductivity of the CR in H^+-ion and the AR in OH^--ions form the current density distribution in the stack is affected. This, in turn, affects the co-ion transport through the anion-exchange membrane and the transport number of co-ions through this membrane will not be constant over the whole range of investigated current densities, as it is assumed in Eq. (2.43). In addition, the changes of the current density affect the concentration polarization and through the Donnan equilibrium cause changes in the ion composition. All above mentioned phenomena do influence both, the ion removal and the co-ion transport into the diluate and contribute to the less favorable results of the CEDI-BM stack as compared to the CEDI-BM-Pro stack.

The results of the experiments with the CEDI-BM-Pro on the other hand demonstrate clearly, that the co-ions transport through the anion-exchange membrane into the diluate can be practically eliminated if the protection compartment is rinsed with a part of the produced diluate.

The use of the diluate to rinse the PC, however, decreases the overall production efficiency of the deionization process and increases the energy consumption. The conductivity of the PC-rinse is comparable with the conductivity of RO-permeate and is still more than 10^4 times lower than the conductivity of the anion-exchange resin in the PC. It is therefore reasonable to use RO-permeate instead of the diluate to rinse the PC and save the produced diluate. This will be shown in Chapter 5.3.4.

5.3.3 Current-voltage curves

Fig. 5.11 shows the effect of the current density on the stack voltage (U) in the above described experiments with the CEDI-BM and the CEDI-BM-Pro.

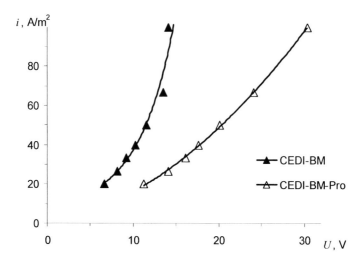

Fig. 5.11. Effect of the current density on the stack voltage

It can be seen from **Fig. 5.11**, that the current density increases with the stack voltage in both stacks, and the slope of the curve becomes steeper at higher current densities. This trend of i versus U curves in the CEDI-BM with and without the PC agrees well with literature data for the CEDI with a mixed-bed [218,219] and also with the i versus U curve of **Fig. 3.30** in Chapter 3. It shows no sign of a limiting current density as in conventional electrodialysis with neutral spacers.

Fig. 5.11 also shows that the stack voltage of the CEDI-BM-Pro is always higher than that of the CEDI-BM at the same current density, because of the additional protection compartment.

In a first approximation, the current density i through the ion-exchange resin of a diluate compartment in a CEDI with separated beds can be calculated by Eq. (2.28) similarly as for electrolyte solutions:

$$i = \frac{\overline{\kappa}}{h}\Delta\phi, \tag{5.12}$$

where h is the diluate compartment thickness, $\overline{\kappa}$ the average conductivity of ion-exchange resin bed, and $\Delta\phi$ the potential difference over the diluate compartment. The value of $\overline{\kappa}$ is in between the conductivity of the resin bed in pure salt form and in the regenerated form, i.e. in form of H^+- and OH^--ions, respectively.

In CEDI, the conductivity of the ion-exchange resins mainly defines the ratio between i und U. The current density increase results in larger amount of generated H^+- and OH^--ions and, subsequently, in a higher fraction of these ions in the ion-exchange resins of the first and second diluate compartment as well as in the respective membranes. Due to the better conductivity of strongly acidic cation-exchange resin in H^+-form and strongly basic anion-exchange resin in OH^--form, compared to any other ionic form, the specific conductivity of the resins increases with the increasing current density, leading to the increasingly steeper slope of the i versus U curve.

5.3.4 Experiments using RO-permeate to rinse the protection compartment

In order to save the diluate used to rinse the protection compartment, the protection compartment was rinsed by RO-permeate, applying different current densities. In these experiments the conditions are the same as in the previous experiments, except the feed temperature, which has risen to 21-22°C due to the season changes.

The measured conductivity of the diluate produced in these experiments is shown in **Fig. 5.12** in comparison with κ_D from the previous experiments where a part of the diluate was used to rinse the PC.

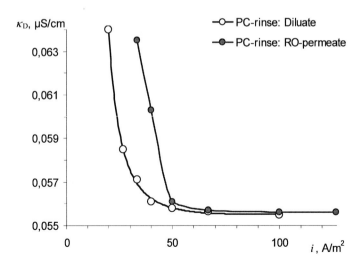

Fig. 5.12. Effect of the current density on the diluate conductivity in CEDI-BM-Pro when diluate or RO-permeate is utilized to rinse the PC (Q_F = 48 L/h; Q_{PC} = 10 L/h)

Fig. 5.12 shows that in both cases κ_D monotonously decreases with increasing current density. Obviously somewhat higher diluate conductivity is obtained when the PC is rinsed with RO-permeate than when it is rinsed with a part of the produced diluate. The difference between the two diluate conductivities decreases however with increasing current density and becomes negligible at current densities above ca. 50 A/m^2. This fact indicates that the influence of cations migration through the AM into the diluate in both experiments is negligible.

The lateral shift between both curves in **Fig. 5.12** is a result of the higher ion concentration of the PC rinse with RO-permeate. It leads to differences in the current density distribution over the stack length, caused by changes of the anions distribution profiles in the protection compartment. Rinsing the protection compartment with RO-permeate introduces the anions from the RO-permeate, which are subsequently exchanged by OH$^-$-ions from the anion-exchange resin. A higher content of mineral anions in the upper (entrance) side of the PC results in a lower conductivity of the AR in this part of the bed, compare to the conductivity of the AR in the lower part of the PC, where the AR is present mostly in the OH$^-$-ion form. Such changes in conductivity of the resin cause changes in the current density distribution in the stack which influence the ions removal and can cause the difference shown in the **Fig. 5.12**.

Like in CEDI-BM, also in CEDI-BM-Pro some Na$^+$-cations from the concentrate compartment will permeate the anion-exchange membrane, here ^{PC}AM. In the protection compartment they will be released into the rinse water, where they couple with some anions, mostly of OH$^-$-form. The dependency of the PC outlet conductivity κ_{PC} upon current density can be calculated as for κ_D using Eqs. (5.8) - (5.11). However Eq. (5.11) for the PC outlet concentration ($^{PC}C_{Na^+}^{OUT}$) needs to be extended to consider the Na$^+$-ion concentration in the PC feed ($^{PC}C_{Na^+}^{IN}$):

$$^{PC}C_{Na^+}^{OUT} = {}^{PC}C_{Na^+}^{IN} + \frac{i \cdot A \cdot \overline{t_{Na^+}^{PC\,AM}}}{F \cdot Q_{PC}}, \tag{5.13}$$

where $\overline{t_{Na^+}^{PC\,AM}}$ is the transport number of Na$^+$-ions through the ^{PC}AM and Q_{PC} – the flow rate of the PC-rinse.

The calculation gives the best correlation with experimental results at high current densities with $\overline{t_{Na^+}^{PC\,AM}} = 0{,}11\ \%$, when diluate is used to rinse the PC, and with

$\overline{{}^{PC}t_{Na^+}^{AM}} = 0{,}32\ \%$ when RO-permeate is used. These values also differ from the co-ion transport number of $\overline{t_{Na^+}^{AM}} \approx 0{,}05\ \%$, found for the anion-exchange membrane in the experiments with CEDI-BM (**Fig. 5.10**).

The differences between transport numbers can be explained by different temperatures, the use of different membrane pieces and somewhat different current distributions as well as pressure differences. The pressure of the diluate in the CEDI-BM was higher because of the higher diluate flow rate compared to the PC rinse. This is probably the reason for the low transport number of $\overline{t_{Na^+}^{AM}} \approx 0{,}05\ \%$.

The calculated dependence fits the experimental data by the integral transport number of co-ions through the ^{PC}AM of $\overline{{}^{PC}t_{Na^+}^{AM}} \approx 0{,}11\%$ as shown in the **Fig. 5.13**. This value is higher than the co-ion transport number found for the AM in the experiments with CEDI-BM ($\overline{t_{Na^+}^{AM}} \approx 0{,}05\ \%$).

5.3.5 Energy consumptions

Fig. 5.14 shows the correlation between the diluate conductivity and the energy (E^V) required to produce 1 m^3 of diluate (Eq. (2.46)) with the CEDI-BM and with the CEDI-BM-Pro, when a part of the diluate or RO-permeate is used to rinse the PC. The two curves reflect the behavior already discussed in connection with **Fig. 5.10**. Due to the limited permselectivity of the anion-exchange membrane, the CEDI-BM diluate conductivity has a minimum at a defined current density and the increase of the current above this value will increase both the diluate conductivity and the energy consumption.

In the CEDI-BM-Pro the diluate conductivity decreases monotonously with increasing energy consumptions. In the range of energy consumptions of $E < 0{,}6\ \mathrm{kW \cdot h/m^3}$, which corresponds to the typical energy consumptions in most commercial electrodeionization modules, the decrease of κ_D is steep and reaches a κ_D of about 0,056 µS/cm. The decreasing of κ_D below 0,056 µS/cm requires a significant increase in energy since the slope of the κ_D - E^V curve is continuously decreasing. It can be assumed that a practical application of the CEDI-BM-Pro for the production of deionized water will be economically reasonable up to a value of 0,6 kW·h/m^3, which corresponds to current densities below 50 A/m^2 to 60 A/m^2 (see **Fig. 5.12**).

178

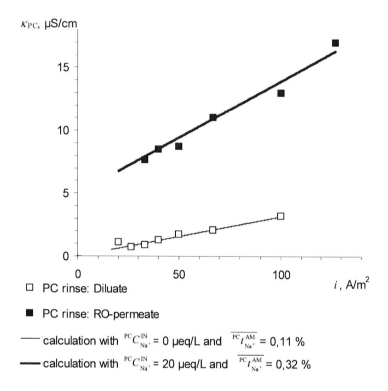

Fig. 5.13. Effect of the current density i on the conductivity κ_{PC} of PC rinse water in CEDI-BM-Pro ($Q_F = 48$ L/h; $Q_{PC} = 10$ L/h)

Interestingly, **Fig. 5.14** also reveals, that the experiments with diluate as PC rinse turn out to be somewhat more energy efficient for the same diluate conductivity than the experiments with RO-permeate. In spite of the 20% lower diluate production rate in the case when 10 L/h of the diluate is used to rinse the PC, the energy consumption for the same diluate conductivity is slightly lower than in the case when RO-permeate is used to rinse the PC because of better deionization performance.

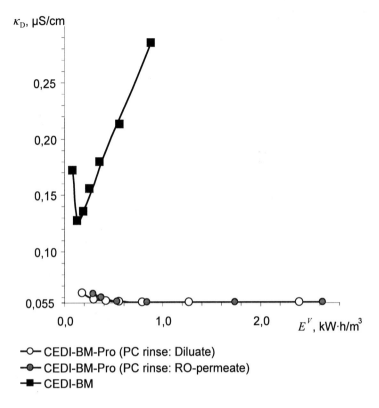

Fig. 5.14. Correlation between the diluate conductivity and the energy consumed to produce 1 m^3 of diluate as product (Q_F = 48 L/h; Q_{PC} = 10 L/h)

5.3.6   Effect of the flow rate in the protection compartment

An increase of the flow rate through the protection compartment leads to a decrease of the concentration of contaminants penetrating through the ^{PC}AM into the PC. It has already been discussed in connection with **Fig. 5.1** that the front of the co-ions (Na$^+$-ions) penetrating the protection compartment (the grey shaded area in the PC in **Fig. 5.1**), should not reach the anion-exchange membrane AM of the AR-diluate compartment. This can be prevented by a higher flow rate, a larger thickness and a shorter length of the protection compartment. The dependency of the diluate conductivity on the flow rate in the PC is shown in **Fig. 5.15**. The flow rate in the PC should be high enough to prevent the co-ions leakage but it is uneconomically to have a too high flow

180

rate, because it reduces the water recovery rate of the module. At current density $i = 50$ A/m^2 this is obviously the case if more than 3 L/h of diluate is used for rinse. Compared to the PC rinse with $Q_{PC} = 10$ L/h of diluate used in the previous experiments, this means that the PC rinse could be reduced to about 6% of the diluate produced at $i = 50$ A/m^2, without loss in diluate conductivity.

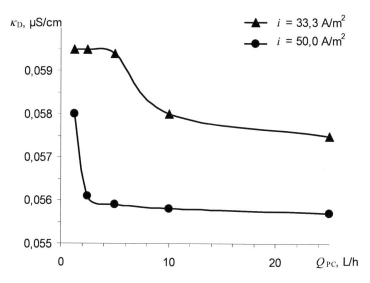

Fig. 5.15. Effect of the flow rate in the protection compartment Q_{PC} on the diluate conductivity κ_D at different current densities ($Q_F = 48$ L/h; part of the diluate is used as PC-rinse)

Fig. 5.16 shows the corresponding change of the outlet conductivity of the rinse flow.

It can be seen from the **Fig. 5.15** and **Fig. 5.16** that, at both tested current densities both, κ_D and κ_{PC} increase with the decrease of Q_{PC}. Even if Q_{PC} is reduced to 1,25 L/h, which accounts for only 2,5% of diluate, κ_D can be maintained below 0,060 µS/cm and 0,058 µS/cm, respectively. The higher conductivity of the PC-rinse at $i = 50$ A/m^2 comparing to $i = 33$ A/m^2 is explained by the higher Na$^+$-ions transport from the concentrate compartment, and the lower diluate conductivity is the result of the better regeneration.

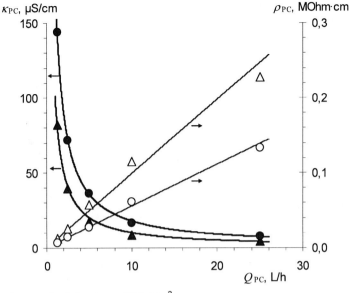

Fig. 5.16. Effect of the flow rate in the protection compartment Q_{PC} on the outlet conductivity κ_{PC} and the resistivity ρ_{PC} of the PC-rinse ($Q_F = 48$ L/h)

At constant current density, the regeneration degree of the ion-exchange resins in the diluate compartments can be considered to be about the same at all rinse velocities. Then the changes in the diluate conductivity with the increasing Q_{PC} are related to the changes of Na^+-ions transport from the PC through the AM into the diluate.

It is evident from Eq. (5.11) that the concentration of Na^+-ions transported from the concentrate compartment through the ^{PC}AM into the PC is inversely proportional to the flow rate of the PC. Thus, in the outlet of the PC rinse the conductivity will decrease hyperbolically and the resistivity ρ_{PC} will increase linearly with increasing PC flow rate as shown in **Fig. 5.16**. The solid lines in **Fig. 5.16** show the conductivity calculated from the Eq. (5.8) using the system of Eqs. (5.11) - (5.10) to calculate ion concentrations. The average transport number of sodium ions was used as fitting parameter, and the best approximation to experimental data was achieved with $\overline{t_{Na^+}^{AM}} = 1,1\%$ for the current density of $i = 33,3$ A/m², and $\overline{t_{Na^+}^{AM}} = 1,3\%$ for the current density of $i = 50$ A/m². The slightly increasing sodium transport through the AM with an increase of current density from 33,3 A/m² to 50 A/m² can be explained by the increase of the electrolyte concentration at the AM surface in the concentrate compartment due to the concentration polarization.

5.3.7   Effect of feed flow rate in the CEDI-BM-Pro

The experiments with different feed flow rates where carried out in the CEDI-BM-Pro stack (**Fig. 5.4***b*) with a part of the diluate used to rinse the PC at a flow rate of $Q_{PC} = 2,5$ L/h and a current density of $i = 50$ A/m² and a feed conductivity of about 2,3 µS/cm. **Fig. 5.17** shows the increase of the diluate conductivity with the feed flow rate.

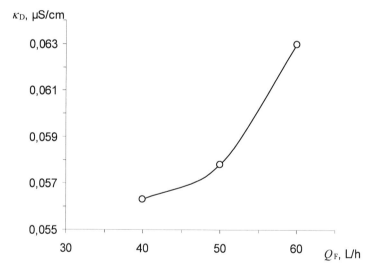

Fig. 5.17. Effect of the feed flow rate on the diluate conductivity at $i = 50$ A/m²; $Q_{PC} = 2,5$ L/h; $\kappa_F = 2,3$ µS/cm

The increasing feed flow rate results in a reduced residence time of the water in the ion-exchange resin beds. The amount of electrolyte species removed by ion-exchange is therefore decreased with increasing flow rate, leading to reduced absorption of weak acids. Thus, with increasing flow rate more ions are breaking through the ion-exchange bed, increasing the diluate conductivity.

Since the feed flow rate Q_F minus the diluate fraction for the PC rinse corresponds to the diluate production, a higher Q_F increases the production capacity of the unit and decreases the investment cost. Optimization of the feed flow rate in a CEDI-BM-Pro configuration is a compromise between the product water quality and factors like investment cost, module footprint, required pressure, etc.

5.3.8 Comparison of different flow directions in the CEDI-BM-Pro stack

Three different combinations of flow directions in the compartments of the CEDI-BM-Pro stack have been tested experimentally. The three combination types are indicated by arrows in the **Table 5.3**. Type-1 relates to the flow directions schematically shown in **Fig. 5.4**b. In Type-2 the flow direction in the PC is reversed, i.e. directed upward. The Type-3 differs from **Fig. 5.4**b in the flow direction through the diluate compartment filled with the AR, i.e. the flow in the AR-filled diluate compartment is upwards, while the flow directions in the CR-filled compartment and in the PC are downwards.

In these experiments a part of the diluate is used to rinse the protection compartment with the flow rate of $Q_{PC} = 2,5$ L/h. The applied current density is $i = 50$ A/m^2, and the other conditions are the same as in previous experiments. The results are summarized in **Table 5.3** and in **Fig. 5.18**.

Table 5.3

Comparison of different flow directions in CEDI-BM-Pro·

Combination of flow directions	Flow direction in compartment			Diluate quality			
	Diluate CR	Diluate AR	PC	κ_D, µS/cm	ρ_D, MOhm·cm	κ_{PC}, µS/cm	U, V
Type-1	↓	↓	↓	0,0578	17,30	83,2	16
Type-2	↓	↓	↑	0,0571	17,51	48,5	16
Type-3	↓	↑	↓	0,0557	17,95	24,4	29

· Flow rate of feed and PC-rinse is 50 L/h and 2,5 L/h, respectively, at current density 50 A/m^2

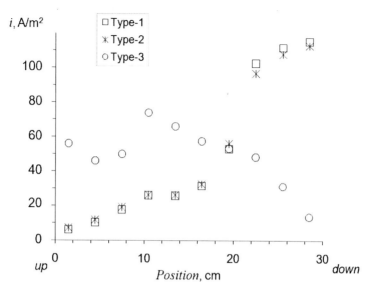

Fig. 5.18. Current density distribution along the anode at different types of flow directions in the CEDI-BM-Pro stack

It can be seen from **Table 5.3** that among the tested flow directions, Type-2 provides lower diluate conductivity than the Type-1, and Type-3 shows the lowest diluate conductivity of all. The same tendency is true also for the conductivities in PC-rinse, which is maximal in the Type-1 and minimal in the Type-3.

It can be seen from the experiments with different PC-rinse flow rates and from the experiments with different flow directions in the stack that at a constant overall current density the diluate conductivity increases with an increase of the conductivity in the PC rinse. This is due to the leakage of co-ions from the PC into the diluate through the AM. It proves that for obtaining the diluate with lowest conductivity, the conductivity in PC-rinse should be as low as possible.

The observed (**Fig. 5.18**) current density distribution is mainly a result of difference in the conductivity of ion-exchange resins being in a regenerated and in a salt form. An estimate of the specific conductivities of the ion-exchange resin and water along the compartments is shown in **Table 5.4**.

Table 5.4

Estimated conductivities of water and ion-exchange resins in the compartments of the

CEDI-BM-Pro stack

	Cation-exchange resin (CR)		Anion-exchange resin (AR)		Protection compartment (PC)	
	entrance	exit	entrance	exit	entrance	exit
Ionic content of water	Feed (RO-permeate)	H^+ and anions from RO-permeate	H^+ and anions from RO-permeate	Diluate (H_2O)	Diluate (H_2O)	NaOH (PC-rinse)
κ_{Water}, μS/cm	2,3	3,0	3,0	0,06	0,06	1 - 3
Form of resin	Na^+	H^+	SO_4^{2-}	OH^-	SO_4^{2-}	OH^-
$\bar{\kappa}_{IER}$, S/m	2,35*	11,1*	4,3**	15,4**	4,3**	15,4**
Conductivity ratio $\dfrac{\bar{\kappa}_{IER}}{\kappa_{Water}}$	10 000	20 000	15 000	2 500 000	700 000	80 000

*Measurements for strongly acidic sulfonated polystyrene-*co*-DVB (8%) CR KU-2-8 (Russia)

**Values from reference [193]

As shown by the estimates in **Table 5.4** the specific conductivity of an ion-exchange resin is some orders of magnitude larger than that of water. Taking into account, that the ion-exchange resin beds form conducting paths for the current and have a volume about twice the volume of water, it can be assumed that the current is conducted almost completely through the ion-exchange resin. This estimate indicates that the specific conductivity of both, cation- and anion-exchange bed, at the inlet of a compartment is few times higher than at the outlet.

Fig. 5.18 shows that the current density distribution for the flow directions of Type-3 is quite different from Type-1 and Type-2. It is evident that in Type-3 the current at the upper side of the anode is higher than at the bottom side and the current density distribution is more even than in Type-1 and Type-2. This behavior is a consequence of the direction of the diluate stream in the AR-filled compartment. It results in the higher conductive OH-form of the AR in the upper (exit) side of the bed, and the lower conductive salt form of the AR in the bottom (entrance) side. In the CR diluate bed the situation is just reversed, resulting in the more even current distribution compared to Type-1 and Type-2.

The fact that only diluate is present at both sides of ^{PC}AM at the diluate exit also applies to Type-2, although both flow directions (of diluate and PC rinse) are now reversed. Therefore Type-2 leads to a lower diluate conductivity than Type-1, although the flow directions in both the CR and the AR compartments and hence the current density distribution are almost identical between Type-1 and -2.

The more even current density distribution in Type-3 (see the **Fig. 5.18**) however results in a substantially higher voltage, required for the same specified current compared to Type-1 and Type-2, since the lines of the current path are longer. Because of the lower voltage drop and energy consumption compared to Type-3, Type-2 is chosen for the further experiments.

5.3.9   Comparison of the unit with protection compartment and with concentrate compartment filled by anion-exchange resin: Influence of the feed conductivity

In his experiments S. Thate found, that a substantial improvement over the conventional CEDI-BM design of **Fig. 4.8** without ion-exchange resins in the concentrate compartment can be achieved, if the concentrate compartment was filled with an anion-exchange resin [193]. Therefore, a module corresponding to the **Fig. 5.5** was assembled, where a unit cell of the CEDI-BM-Pro (with protection compartment) and the flow Type-2 was tested against a unit cell with the concentrate compartment CC filled with an anion-exchange resin under the same operating conditions. Different feed conductivities and current densities have been applied. In order to adjust the conductivity of the RO-permeate used as feed, the flow rate of the RO-retentate in the RO-set-up (**Fig. 3.19**) was regulated. A decrease of the RO-reject flow results in an increase of RO-recovery rate and permeate conductivity.

Measured values of diluate conductivity at different feed conductivities (κ_F) and current densities are shown in **Fig. 5.19**. It can be seen, that the obtained conductivity of the diluate in most experiments is lower than 0,065 μS/cm ($\rho_D > 15$ μS/cm) for both units. This demonstrates the ability to obtain high-purity water with CEDI-BM using either a PC or a concentrate compartment filled with an AR.

The calculated values of the current efficiency (ε) for the removal of Na^+-ions from the feed, are also presented in **Fig. 5.19** and **Fig. 5.21**. It should be notes that at both tested current densities a nearly complete removal of ions takes place and the difference between Na^+-ion concentrations in the feed and in the diluate is approximately equal to the sodium feed concentration. Thus, the current efficiency depends upon the feed flow rate, the Na^+-ion concentration in the feed and on the current strength. Therefore, ε increases linearly with the feed concentration (conductivity) and decreases hyperbolically with the current (Eq. (2.47)).

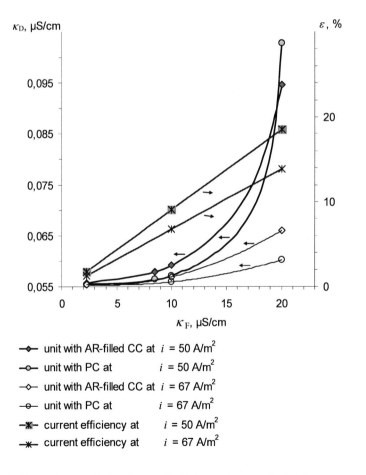

Fig. 5.19. Dependence of diluate conductivity and current efficiency upon feed conductivity and current density at flow rate of 2,5 L/h through PC and AR-filled CC

Since both investigated units show almost complete ion removal, the dependency of ε upon κ_F and i is practically the same for both stacks; it is shown in **Fig. 5.19** and **Fig. 5.21** as only one line for each current density.

The increase of κ_D with increasing κ_F, or with decreasing i, indicates a reduction in the regeneration degree of the ion-exchange resins. First, the reduction of the AR regeneration becomes apparent. It leads to a breakthrough of some weak acids through the AR bed and a subsequent increase of the diluate conductivity. In the experiment with the lowest tested current density and the highest feed conductivity a breakthrough of contaminants not removed in the diluate compartments becomes significant and leads to a rapid increase of diluate conductivity over feed conductivity.

In most experiments the diluate conductivity of the unit with a protection compartment is slightly lower than that of the unit containing an AR-filled concentrate compartment. This fact indicates the better elimination of co-ions leakage through the ^{PC}AM in the CEDI-BM-Pro unit. Only in the experiment with $i = 50$ A/m^2 and $\kappa_F = 20$ µS/cm the diluate conductivity in the unit with the PC is higher. The high values of diluate conductivity in this experiment indicate the dominating influence of insufficient ion removal in the two diluate beds, while the co-ions leakage from the concentrate seems to be not decisive for the diluate conductivity. Under these conditions small differences in the current or the flow distribution in the investigated units could affect the performance stronger than the differences in co-ions transport.

The influence of co-ions transport in both units becomes more evident considering the conductivities of the rinse water from the PC, κ_{PC} and from the concentrate compartment, κ_{CC}. The dependence of κ_{PC} and κ_{CC} upon the feed conductivity κ_F is shown in the **Fig. 5.20**.

It is remarkable, that at the experimental conditions the conductivity of the concentrate compartment rinse is about one order of magnitude higher than that of the PC rinse. As a result of a relatively high electrolyte concentration in the concentrate compartment rinse more cations can reach the surface of the AM and penetrate into the diluate, leading to the observed differences in κ_D (**Fig. 5.19**).

If the deionization degree approaches 100 % almost all ions from the feed are collected in the concentrate compartment and rinsed away, whereas the ions removed with the PC rinse are negligible (see κ_{PC} over κ_F). The excess current will be conducted by H^+- and OH^--ions, which will recombine in the concentrate compartment. Thus, in both configurations the conductivity of the concentrate compartment rinse is directly proportional to the feed conductivity (**Fig. 5.20**) and the slope of the κ_{CC} vs. κ_F dependency corresponds to the ratio of the feed flow rate to the flow rate in the concentrate compartment.

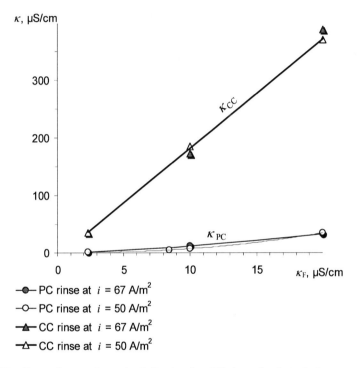

Fig. 5.20. Dependence of conductivity in the PC-rinse (κ_{PC}) and the concentrate compartment rinse (κ_{CC}) on the feed conductivity at flow rate of 2,5 L/h in PC- and CC-rinse

5.3.10 Influence of the current density

The values of the diluate conductivity obtained in experiments with a feed conductivity of 20 µS/cm and different current densities are shown in **Fig. 5.21**. The figure shows an exponential decrease of the diluate conductivity with increasing current density, which is due to a better regeneration of the ion-exchange resins, while co-ions leakage is almost eliminated.

The rate of decrease in the diluate conductivity is quite high in the range of low current densities, where the removal of strongly dissociated electrolytes occurs. At higher current densities the decrease of κ_D with i levels off since it corresponds to the removal of the remaining weakly dissociated electrolytes. In this range of i the diluate from the unit with PC has a somewhat lower conductivity than from the unit with AR-filled concentrate compartment, but with increasing current density both κ_D values approach each other.

In practical applications the exponential decrease of the $\kappa_D - i$ dependency allows to adjust the current corresponding to the required diluate quality.

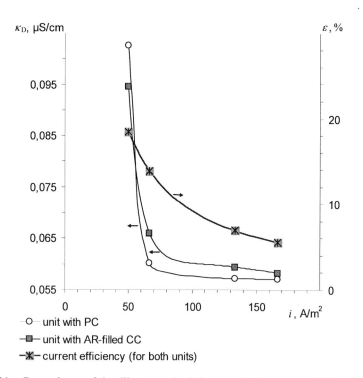

Fig. 5.21. Dependence of the diluate conductivity κ_D and the current efficiency ε upon current the density i at a feed conductivity $\kappa_F = 20$ µS/cm

The decrease of κ_D with i is accompanied by a decreasing current efficiency ε (**Fig. 5.21**). The obtained values of current efficiency are well in the range of current efficiencies usually claimed for typical commercial CEDI modules.

Both investigated units show almost complete ion removal. The dependence of ε upon the κ_F and i is therefore practically the same for both units as already shown in **Fig. 5.19**.

In **Fig. 5.22** the dependence of the water conductivity in the PC-rinse and in the concentrate compartment rinse on the current density is shown. As expected, the conductivity in the PC rinse increases with the current, because of the increasing amount of co-ions transported through the ^{PC}AM into the protection compartment.

The conductivity of the concentrate compartment rinse remains practically constant in the investigated range of current densities due to the almost complete ion removal as explained above. Therefore at high current densities the electrolyte concentration in the outlet of the concentrate compartment, κ_{CC} is not effected by the current and depends on the feed concentration and on the flow rates of the feed and the concentrate compartment rinse.

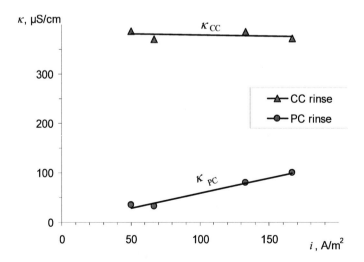

Fig. 5.22. Dependence of the conductivities in the rinse water of the protection compartment (κ_{PC}) and of the concentrate compartment (κ_{CC}) upon the current density at the $\kappa_F = 20$ µS/cm. $Q_{PC} = Q_{CC} = 2,5$ L/h

5.3.11   Si removal

The total Si concentration in the feed and the diluate has been measured by Hager + Elsässer GmbH (Germany) using atomic absorption spectrophotometry with graphite furnace. The results of the experiment with a feed conductivity of 5,5 µS/cm and a current density of 50 A/m^2 are summarized in **Table 5.5**.

Table 5.5

Si concentration and conductivity in feed and in diluate

Unit	Feed			$i,$	Diluate			Si removal,
	$\kappa,$	$\rho,$	Si,		$\kappa,$	$\rho,$	Si,	
	µS/cm	MOhm·cm	ppb	A/m^2	µS/cm	MOhm·cm	ppb	%
AR-filled PC	5,5	0,2	285	50	0,0556	18,0	0,64	99,78
AR-filled CC					0,0563	17,8	2,7	99,05
Mixed-bed EDI [220]	12,4	0,1	92	Not specified	0,056	17,8	8	91

Table 5.5 shows that both units provide very good removal of Si. The diluate from the unit with PC contains Si concentrations below 1 ppb, which is lower than in the unit with the AR-filled concentrate compartment. At higher current density and (or) lower Si concentrations in the feed the Si concentration in diluate can be lowered further. For example, using a feed with 135 ppb of Si and a $\kappa_F = 2,3$ µS/cm, also the unit with AR-filled concentrate compartment, operating at $i = 133$ A/m^2, produces a diluate with a Si concentration of 0,24 ppb and a $\kappa_D = 0,0553$ µS/cm.

The Si removal obtained in both units is significantly higher than in commercial CEDI stacks with mixed-bed, as shown in the **Table 5.5**. This can be explained by advantages of separated beds that are used in the CEDI-BM.

5.3.12   Coupling of the CEDI-BM with reverse osmosis

It is obvious from **Fig. 5.21** that the CEDI-BM with protection compartment or with AR-filled concentrate compartment can deionize RO-permeate of relatively high conductivity ($\kappa_F = 20$ µS/cm) to the level of pure water (15 - 17,5) MOhm·cm. The current efficiency increases with the conductivity of the RO-permeate used as feed water. Moreover, the production of RO-permeate with higher conductivity allows higher recovery rates of the RO-unit, reducing the volume of RO-retentate to be disposed. Reduction of the RO-retentate is sometimes very important in big plants or closed systems.

The dependence of the RO-permeate conductivity on the ratio of permeate to retentate flow, Q_P/Q_R, for the RO-unit used in the experiments is depicted in **Fig. 5.23**. It shows a linear increase of the permeate conductivity (κ_F) with Q_P/Q_R. The increase of Q_P/Q_R also leads to an increase of the RO recovery rate (Δ_{RO}). It is therefore reasonable to operate the RO-unit in the middle of the grey shaded area, where the RO-recovery rate is already high but its conductivity is still low enough to be efficiently deionized by CEDI-BM. This is particularly true for applications, where the required diluate resistivity of (15 - 17) MOhm·cm is sufficient. In this case a high RO-recovery rate (about 80 % - 90 %) can be well combined with a high current efficiency in the CEDI-BM.

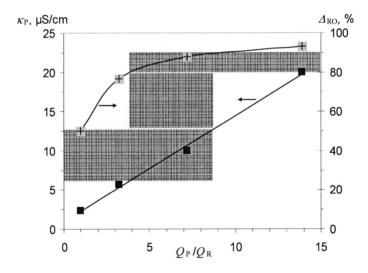

Fig. 5.23. Dependence of the conductivity of RO-permeate, κ_P, and the RO-recovery rate Δ_{RO} upon the ratio between RO-permeate and retentate flow rates Q_P/Q_R, measured for a conductivity of softened tap-water as RO-feed of 335 µS/cm

5.4 CEDI-BM summary and outlook

The review of different known CEDI technology concepts, based on literature data and considering ionic transport phenomena, leads to the choice of the CEDI with separated beds of cation- and anion-exchange resins, and particularly to the CEDI-BM (**Fig. 4.8**), as the most advantageous concept for effective deionization. However, because of a limited permselectivity of anion-exchange membranes, some co-ion leakage occurs, which disqualifies this approach for ultrapure water production.

In the experiments with the original concept of CEDI-BM, the influence of the anion-exchange membrane permselectivity on the conductivity of the produced diluate has been investigated in more details and data published in the literature have been confirmed and rationalized. The experiments with variations of the pressure and flow rate in the concentrate compartment as well as the increase of the diluate conductivity with the current density in the range of high current densities (**Fig. 5.10**) have clearly shown, that membrane permselectivity is the main limitation of the original CEDI-BM concept. An optimized current density is required to minimize the effect of co-ion transport into the diluate. Experimentally determined relations between diluate conductivity, flow rate and pressure in the concentrate compartment are in good agreement with simplified mathematical approximations. In these approximations the concentration polarization in the diffusion layer of the membrane facing the concentrate compartment filled with neutral spacer is identified as the main reason which also contributes to the amount of co-ion leakage.

To solve the problems associated with limited membrane permselectivity, an additional protection compartment between the last diluate and the concentrate compartment has been proposed. This compartment should be filled with an anion-exchanger and separated from the adjacent concentrate compartment by an anion-exchange membrane. If this protection compartment is rinsed with a small fraction of the diluate, the leakage of cations from the concentrate over the protection compartment into the diluate can be almost completely prevented.

This new concept has been compared with a concept, where the concentrate compartment was filled with anion-exchange resin. The feasibility of this solution has already been studied in short-term tests by S. Thate [193]. Furthermore, it has recently become good practice of CEDI manufacturers to use concentrate compartments filled with an ion-exchange resin in the CEDI modules with mixed-bed or layered bed concepts.

The two above modifications of CEDI-BM units have been combined into a single stack in order to compare them under identical experimental conditions. Both tested concepts confirm significant improvements of the purification performance in terms of the achieved diluate resistivity, compared to original CEDI-BM concept. Because of high conductivity of resins compare to water, in both CEDI-BM modifications with anion-exchange resin filled protection or concentrate compartment the current in this compartment, adjacent to diluate compartment, is conducted to big extent through the membrane/resin and not through the membrane/solution interphase. That prevents concentration polarization, i.e. concentration increase on membrane surface in either protection or concentrate compartment and significantly reduces following penetration of contaminants through the anion-exchange membrane into diluate.

The variation of different parameters such as current, flow rates and flow directions in different compartments have been tested. The relation between operating parameters and purification performance has been established and the optimum range of operating parameters has been identified.

Experiments with different current densities in the CEDI-BM with protection compartment show that the diluate conductivity decreases monotonously with increasing current density since the co-ions transport into the diluate is eliminated. The additional compartment, however, leads to an increase of the energy consumption. Concerning the relation of diluate conductivity, κ_D over current density i, two current regimes can be distinguished. At lower current densities a strong decrease of the conductivity with i occurs due to the removal of the strongly and the majority of the weakly dissociated electrolytes. If the current is increased further, the decrease of κ_D is weaker. This weak decrease is due to the removal of the residual weakly dissociated electrolytes. In both of these current regimes the removal of contaminants is considered to occur by the mechanism of ion-exchange with electrochemical regeneration of resins.

The regime of weak decrease of κ_D corresponds to a strong decrease of the current efficiency and an increase of the operating costs. In a practical application the current density should therefore be limited in order to optimize energy consumption and water production. The use of high current densities in a module is also of disadvantage for the following reasons:

- increasing gas formation at the electrodes with consequent electric shadow effect of the gas-bubbles, higher flow velocity in the upper (exit) part of the electrode compartments, flow channeling in electrode compartments and risk of explosion;
- faster ageing of ion-exchange materials resulting in a long-term growth of electrical resistance of the stack;
- rise of the stack voltage to values which require additional safety precautions.

The choice of the appropriate current strength for a particular application also requires to balance a reduced product water quality at lower operating cost against the additional costs of removing residual contaminants by a final polishing step, e.g. by a mixed-bed ion-exchange.

The remaining difference between the experimentally determined diluate conductivity and the theoretical conductivity of ultrapure water can generally be attributed to the residual of weak acids and of minor contaminations through the membranes in CEDI-BM. Comparing modifications with protection compartment and with AR-filled concentrate compartment at the same operating conditions the removal of weakly dissociated acids can be affected by current density distribution, while the prevention of contaminants penetrating into dilute is also affected by the concentration of contaminants in different parts of a compartment neighbor to diluate.

The always lower diluate conductivity obtained in CEDI-BM unit with protection compartment compare to the unit with AR-filled concentrate compartment indicates that, probably, some minor effect of back-diffusion and co-ions leakage into the diluate from the AR-filled concentrate compartment is possible, but a different current density distribution can be the reason as well.

The experimental results have shown that the problem of non-ideal membrane permselectivity can be solved satisfactorily in the CEDI with a protection compartment or an AR-filled concentrate compartment. Nevertheless, other phenomena like distribution of current density, impacting the product water quality should be addressed for further improvements.

The experimental comparison of the CEDI-BM unit with a PC and with an AR-filled concentrate compartment has been carried out at different feed conductivities and current densities. The experiments show that the strong influence of the co-ions leakage into the diluate can be eliminated by using a protection compartment or an AR-filled concentrate compartment. A reliable deionization with both units is possible, while the diluate conductivity and the Si concentration in the unit with the protection compartment were slightly lower at most tested experimental conditions. The difference in the diluate conductivities between the units is related to the higher electrolyte concentration in the CC-rinse water compared to the PC-rinse.

The choice of either using a protection compartment or an AR-filled concentrate compartment with a part of the diluate or of the RO-permeate to rinse the compartment adjacent to the diluate compartment, depends on the specific practical application, such as the required quality of product water, the used feed water, the acceptable system recovery rate, the investment and operating costs, etc.

The CEDI-BM with PC was studied experimentally in more detail and the effect of the current density, the flow rate in PC and in diluate compartment as well as different types of flow directions has been investigated. The conductivity of the diluate decreases with an increase of the current density or the flow rate in the PC, and a resistivity of $\rho_D \approx 18$ MOhm·cm can be achieved at high current densities.

In the tested range of flow rates of the CEDI-BM-Pro the diluate conductivity increases monotonously with flow rate. This can be explained by the limited kinetics of ion-exchange and the breakthrough of weakly dissociated species through the bed, even if co-ion penetration into the diluate can be eliminated. If the co-ion transport into the diluate would be dominant, like in the original CEDI-BM, a minimum conductivity at a certain flow rate can be observed, similarly to the minimum in the $\kappa_D - v$ dependence shown in **Fig. 4.3***b*.

The fact, that there are no weakly dissociated bases and that Na^+-cations in the CR are removed easier then acids in the AR, leads to the conclusion that the ion-exchange in the cation-exchange bed is not decisive for the overall deionization. Cation-exchange resins with larger bead sizes than anion-exchange resins can therefore be used to decrease the total pressure drop in a stack, or thinner compartments with cation-exchange resin could be used.

The use of a protection compartment, especially when it is rinsed with a part of the diluate, prevents the negative influence of co-ions transport on the diluate quality and cheaper ion-exchange membranes with lower permselectivity can be applied in CEDI-BM-Pro, in order to decrease the material costs.

It is possible to use both RO-permeate and CEDI-diluate to rinse the PC, but using the diluate results in slightly better product water quality. Studies of using a part of the diluate to rinse the PC at different flow rates show that only 2,5% of the diluate produced is sufficient to rinse the PC without a significant negative impact on product water quality.

The tests with three different combinations of flow directions in the PC and in both diluate compartments at the same current density show, that the flow direction in CEDI-BM can have a strong effect on the required voltage, or on the module resistance. The diluate conductivities obtained in the tested combinations are also slightly different. It was shown that the above mentioned differences are linked to the current density distribution along the stack. Thus, arranging the flows in the cation- and the anion-exchange resin compartments in opposite directions (flow Type-3 in the **Table 5.3**) results in a more uniform current density distribution and a remarkably lower diluate conductivity, but at the expense of a significant increase of the voltage required for a given current. Therefore subsequent experiments have been performed with identical flow direction in the compartments with cation- and anion-exchange resins, but with a flow scheme, where the diluate leaves and the PC-rinse enters the corresponding compartment from the same side of the stack (flow Type-2).

The fact that the lowest diluate conductivity is obtained with the more uniform current density distribution of flow Type-3 should be used for further improvements in deionization performance. In particular it should be checked, whether the energy consumption for a specified diluate conductivity is lower for Type-2 or Type-3.

In tested modifications of CEDI-BM the removal of residuals of weakly dissociated acids for a given ion-exchange materials and size of the compartments can be improved changing the current density distribution, e.g. by selecting flow directions in the compartments. The complete prevention of contaminants from the neighbor compartment penetrating through a membrane into the diluate will require taking into account different parameters in this neighbor compartment, like concentration of contaminants, current density distribution, flow velocity, pressure, etc. Also possible contaminations penetrating the bipolar membrane in the area close to the diluate outlet should be verified. A detailed investigation of such minor effects, which confine the production of diluate with resistivity 18,2 MOhm·cm, is a very challenging task that would require assurance of constant specified experimental conditions during long period of tests and available online-analysis of contaminants like ions, silicium, boron and organic matters in trace level. Results of such investigations, however, could be of a high importance, because an open possibility to reliably produce ultrapure water in electrodeionization can bring a big advantageous simplification for water purification chain in several industries where further polishing step is currently applied downstream the electrodeionization.

A more equal current density distribution than in experiments present in this work could also be achieved by placing a layer of resin with lower conductivity close to the outlet part of a compartment, while a resin with higher conductivity is placed in the inlet part of the compartment. A standard strong-base AR in the inlet part of the diluate compartment and a boron selective resin of Type-II, which has typically a higher resistivity, in the outlet part of compartment, could improve the deionization performance of a system, especially regarding the boron removal [221].

For better current density distribution the AR diluate compartment can also be loaded with a mixture of two resin types, so that the higher conductive AR of Type-I is placed at the compartment inlet and the less conductive AR of Type-II is placed at the compartment outlet, while a gradient of intermediate proportions of resins in the mixture is present in between.

The tests with variation of the feed water conductivity show an increase of κ_D with κ_F. Usually, in practical applications of CEDI, the feed water quality is related to the local water supply, but it can be adjusted by a pretreatment, like reverse osmosis (RO), degassing, UV-oxidation, etc. An increase in the RO-recovery rate results in an increase in the permeate conductivity and will require operation of a downstream placed CEDI module at a higher current density or at lower flow rates. Designing water purification systems, both the RO and the CEDI should be considered and optimized together, with the goal of minimum waste and energy consumption.

Coupling of a CEDI unit with separated beds, i.e. CEDI-BM, with degasser could have additional features. Since water leaving the bed with cation-exchange resin is acidified the inorganic carbon is present there almost completely in non-dissociated CO_2

form. Thus, installing a degasser between the beds of cation- and anion-exchange resin will lead to more efficient removal of CO_2, that can be advantageous by treatment of waters with high content of inorganic carbon.

It should be noted that in this study only water free of hardness ions was studied as feed, while in practical CEDI applications this is not always the case. Many CEDI modules are expected to accept a certain level of hardness in the feed water, which is typically below 1 ppm (as $CaCO_3$) of total hardness. The main risk related to the presence of hardness ions in the feed water is the scaling, possible at higher concentrations and pH-values, with a subsequent increase of the voltage or decrease of current and a decline in the product water quality.

Considering the concept of the CEDI-BM-Pro of **Fig. 5.1** or the concept of the CEDI-BM with an AR-filled concentrate compartment presented in **Fig. 5.2**, one can see that formation of scaling in the diluate compartments is rather impossible, because the feed water passing the first cation-exchange resin compartment will be increasingly acidic. The most sensitive area for possible scaling formation in these concepts is located at the anion-exchange membrane, facing the concentrate compartment, especially in the part close to the diluate outlet, where OH^--ions from the regenerated anion-exchange resin penetrate through the anion-exchange membrane. Nevertheless the scaling risk can be reduced, selecting proper thickness of, and flow directions through, the compartments.

For example, in CEDI-BM with AR-filled concentrate compartment it could be preferable to change the flow directions depicted in **Fig. 5.2** as follows: the CR diluate – downwards, the AR diluate – upwards and the AR-filled concentrate compartment – upwards. In this case the CC-rinse stream entering the concentrate compartment from the bottom will receive first mainly H^+-cations migrating from the more regenerated CR and salt anions migrating from the less regenerated AR in the corresponding diluate compartment. Thus, the water will be slightly acidified. This can prevent a pH increase in the following part close to the concentrate compartment outlet, where mostly OH^--ions migrate from the regenerated AR of diluate compartment. On the other hand, this flow scheme has a disadvantage that the PC rinse with the highest concentration leaves the stack at the same side as the diluate. This discussion shows that the right alternative should be taken with respect to the requirements of the specific task.

One reason for filling the concentrate compartment with an anion-exchange resin (**Fig. 5.2**) is also to reduce the scaling risk. Since the release of OH^--ions into the rinse water of the concentrate compartment will then occur through the high surface of the anion-exchange bed in the concentrate compartment, a local increase of pH at the surface of the anion-exchange membrane (and beads) will not be significant, compared to a concentrate compartment filled with a neutral spacer or a cation-exchange resin.

For the further improvement it is possible to use a portion of the acidic water from the outlet of the cation-exchange resin bed to rinse PC or CC. Since it is easier to remove all cations in the cation-exchange resin bed than the residuals of weakly dissociated acids in the subsequent anion-exchange resin bed, a higher flow rate through the cation-exchange resin could be acceptable. This allows taking some of the effluent from the CR bed to rinse the PC or CC, without affecting the overall productivity. Since the cations are already removed, the acidic pH of the rinse reduces the scaling risk in the PC or CC.

The use of innovative materials, like ion-exchange spacers, fibers, textiles or porous blocks can also help solving some problems related to the utilization of conventional ion-exchange resin beds, since loading and stabilization of the resin beads into the compartments of a module requires special techniques and is time consuming. The volume changes of ion-exchange resin beads in different forms can put some pressure on the membranes and frames, or can create cavities free of resin beads in the bed. This can result in less efficient contact between the membrane surface and the ion-exchange beads, in flow channeling, membrane deformation, etc. A use of flow-through ion-exchange materials prepared as blocks filling the compartments of a CEDI module completely could simplify the manufacturing procedure and improve the quality.

From the purification point of view the use of ion-exchange fibers or porous blocks with a developed specific surface area can also improve the kinetics of ion-exchange, which would reduce the required length of the module compartments, or it would alternatively allow treating water at higher flow rates or with higher electrolytes concentration.

Creation and utilization of self-assemblies from ion-exchange membrane with a dense structure and an ion-exchange block with a porous structure as one unit, e.g. by phase separation from a solution of an ion-exchange polymer, could be another progressive step in the development of electromembrane devices, like CEDI.

Appendix 1

Properties of water at 25°C and 101,3 kPa

Property	Symbol	Value	Unit
Molecular weight	MW	0,01802	kg/mol
Boiling point	T_b	273	K
Melting point	T_m	373	K
Concentration	C	55,5	mol/L
Density	ρ	997	kg/m^3
Dynamic viscosity	η	$8{,}90 \cdot 10^{-4}$	Pa·s
Prandtl Number	Pr	6,10	-
Thermal conductivity	λ	0,610	W/(m·K)
Thermal expansion coefficient	α	$2{,}56 \cdot 10^{-4}$	K^{-1}
Heat capacity	C_p	4181	J/(kg·K)
Relative permittivity (dielectric constant)	ε_r	78,3	-
Specific electric conductivity	κ	0,0550	µS/cm
Molecular size (van der Waals diameter)	d	0,28	nm
Ionic product	K_W	$1{,}008 \cdot 10^{-14}$	(mol/L)^2
Polarizability	α	$1{,}62 \cdot 10^{-40}$	F·m^2

Appendix 2

Classifications of water by salt content

United States Geological Survey		Wikipedia	
	TDS, mg/L		*TDS*, mg/L
Fresh water	< 1000	Fresh water	< 500
Slightly saline	1000 – 3000	Brackish water	500 – 30000
Moderately saline	3000 – 10000	Saline water	30000 – 50000
Highly saline	10000 – 35000 or higher	Brine	Over 50000

Appendix 3

Quality standards for high purity waters

(Maximal allowed values for some critical parameters)

Parameters of ultrapure water for the most critical processes
in wafer environmental contamination control - ITRS 2006
(International technology roadmap for semiconductors) [222]

	κ, μS/cm	TOC^*, ppb	Si (as SiO_2), ppb	Critical metals[†], ppt	Other critical ions[†], ppt	O_2, ppb	Application
UPW	0,055 ρ_{min}= 18,2 MOhm·cm	1	0,5	1	50	10	semiconductors

Standard specification for Reagent Water ASTM D 1193-99[*]
(American Society for Testing and Materials) [223,224]

	κ, μS/cm	TOC, ppb	Si (as SiO_2), ppb	Na^+, ppb	Cl^-, ppb
Type I	0,056 ρ_{min}= 18 MOhm·cm	50	3	1	1
Type II	1	50	3	5	5
Type III	0,25	200	500	10	10
Type IV	5	-	-	50	50

Specification of Reagent Water in the Clinical Laboratory. NCCLS[*][223]
(Clinical and Laboratory Standards Institute)

	κ, μS/cm	TOC, ppb	Si (as SiO_2), ppb	$TDS^{‡}$, ppm	pH
Type I	0,1	0,05	50	0,1	-
Type II	1	0,2	100	1	-
Type III	10	1	1000	5	5,0 - 8,0

---

[*] *TOC*: Total Organic Carbon

[†] Critical metals and ions include: Al, As, Ba, Ca, Co, Cu, Cr, Fe, K, Li, Mg, Mn, Na, Ni, Pb, Sn, Ti, Zn.

[‡] *TDS*: Total Dissolved Solids

Specification for water for laboratory use ISO 3696
(International Organization for Standardization 1987) [223]

	κ, µS/cm	UV_{254}[*], cm^{-1}	Si (as SiO$_2$), ppb	TDS[†], ppm	pH	O$_2$, ppm	Application[‡]
Grade 1	0,1	0,001	10	-	-	-	precise analytic, e.g. HPLC
Grade 2	1,0	0,01	20	1	-	0,08	sensitive analytic, e.g. AAS
Grade 3	5,0	-	-	2	5,0 - 7,5	0,4	most laboratory work, reagent solutions

Pharmacopoeia requirements to purified water [223]

	κ, µS/cm	TOC, ppb	Heavy metals, ppm	NO$_3^-$, ppm
EP	4,3 (at 20°C)	500	0,1	0,2
USP	1,3 (at 25°C)	500	-	-

Water quality classification adapted from British Water [152]

	ρ, MOhm·cm	TOC, ppb	Si (as SiO$_2$), ppb[§]	TDS, ppm	Microorganisms, cfu/mL[**]
Deionized water	0,05	-	0,5	10	-
Purified water	0,2	-	0,1	1	10
Apyrogenic water	0,2	-	0,1	1	1
High purity water	10	-	0,02	0,5	1
Ultrapure water	18	0,05	0,002	0,005	1

Absorbance at 254 nm in 1 cm cell is used to characterize TOC level, which can be estimated with following correlation: $TOC = 0,56 + 30,1 \cdot UV_{254}$ [225].

---

[*] Absorbance at 254 nm in 1 cm cell is used to characterize TOC level.

[†] Residue after evaporation on heating at 110°C.

[‡] Not applicable to organic trace analysis, analysis of surface-active agents and biological or medical analysis.

[§] Reactive silica

[**] cfu – colony forming unit

Appendix 4

Leading CEDI module manufacturers*

Name	Web-site	Module type	Filling of diluate compartments
Christ (part of GLV)	http://www.christwater.com	Spiral wound	Mixed-bed
Dow (former Omexcell)	http://www.dowwaterandprocess.com	Spiral wound	Mixed-bed
GE (former E-Cell and Ionics)	http://www.gewater.com	Plate-and-frame	Mixed-bed
Ionpure (part of Siemens)	http://www.ionpure.com	Plate-and-frame	Layered bed
Merck Millipore	http://www.millipore.com	Plate-and-frame	Mixed-bed
Nippon Rensui	http://www.rensui.co.jp	Plate-and-frame	Grafted materials
Siemens (former SG Water)	http://www.sgwater.de	Plate-and-frame	Separated beds
Snow Pure (former Electropure)	http://www.snowpure.com	Plate-and-frame	Mixed-bed

* Several other companies manufacturing electrodeionization modules exist, while their size and influence on the CEDI-market is small or/and information about their CEDI modules isn't widely open in English: Ebara Engineering Service (Japan), Organo (Japan), Kurita Water Industries (Japan), Mega (Czech Republic), Elga (UK) part of Veolia Water, HAJI (South Korea), Canpure (China), Shandong Zhaojin Motian Company Limited (China), Guangzhou crystalline resource desalination of seawater and water treatment (China), Mahir Technologies (India), etc.

References

1. Baker R.W. Membrane technology and applications. 538 p. John Wiley & Sons, Ltd. 2004. ISBN: 0-470-85445-6.
2. Chaplin M. Water structure and science. http://www.lsbu.ac.uk/water/index2.html
3. Martin F., Zipse H. Charge distribution in the water molecule - A comparison of methods. J. Comput. Chem. V. 26 (2005) pp. 97-105.
4. Császár A.G., Czakó G., Furtenbacher T., Tennyson J., Szalay V., Shirin S.V., Zobov N.F., Polyansky O.L. On equilibrium structures of the water molecule. Journal of Chemical Physics. V. 122 № 21 (2005) pp. 214305.1-214305.10.
5. Hasted J.B. Liquid water: Dielectric properties // in: Water A comprehensive treatise, V. 1, Ed. F. Franks (Plenum Press, New York, 1972) pp. 255-309.
6. Silvestrelli P.L., Parrinello M. Structural, electronic, and bonding properties of liquid water from first principles. J. Chem. Phys. V. 111 (1999) pp. 3572-3580.
7. Markovitch O., Agmon N. Structure and energetics of the hydronium hydration shells. The Journal of Physical Chemistry A. V. 111 № 12 (2007) pp. 2253 – 2256.
8. IUPAC Compendium of Chemical Terminology. http://www.iupac.org/publications/compendium/
9. de Grotthuss C.J.T. Sur la décomposition de l'eau et des corps qu'elle tient en dissolution à l'aide de l'électricité galvanique. In: Ann. Chim. (Paris). V. 58 (1806) pp. 54-73.
10. Debye P., Hückel E. The theory of electrolytes. I. Lowering of freezing point and related phenomena. Physik. Z. V. 24 (1923) pp. 185-206.
11. The collected works of Lars Onsager (With Commentary). 1088 p. World Scientific Series in 20th Century Physics. Vol. 17. Edited by Per Chr Hemmer, Helge Holden & Signe Kjelstrup Ratkje (Norges tekniske høgskole, Trondheim, Norway). 1996. ISBN 981-02-2563-6.
12. Damaskin B.B., Petrii O.A., Tsirlina G.A. Electrochemistry. 624 p. Moscow: Khimiya. 2001. (in Russian)
13. Bevilacqua A.C. Advances in resistivity instrumentation for UPW systems of the future. Semiconductor Pure Water and Chemicals Conference. March 2-5, 1998. 16 p.
14. Nora C., Mabic S., Darbouret D. A theoretical approach to measuring pH and conductivity in high-purity water. Ultrapure Water. V. 19 № 8 (October 2002) pp. 56-61.
15. Schmidt-Nawrot J. Reinstwasser mittels Membranentgasung. Chemicalienfreie Entfernung von Kohlendioxid mit Druckluft. CHEManager 9/2006 p. 49.
16. Temperaturkompensation bei konduktometrischen Messungen. Application report № 53022. WTW.
17. Rhoades J.D., Kandiah A., Mashali A.M. The use of saline waters for crop production. FAO irrigation and drainage Paper 48. Food and agriculture organization of the united nations. Rome, 1992.
18. Kenneth K. Tanji, Neeltje C. Kielen Agricultural drainage water management in arid and semi-arid areas. FAO irrigation and drainage paper 61. Food and agriculture organization of the united nations. Rome, 2002.
19. Ayers R.S. and Westcot D.W. Water quality for agriculture. Irrigation and Drainage Food and agriculture organization of the united nations. (1985) Paper 29 Rev. 1. FAO, Rome. 174 p.
20. ASTM D 5127, "Standard guide for ultra pure water used in the electronics and semiconductor industry," ASTM International.
21. Breedis J. Method of polymerization. US 2326326. Oct. 10, 1943.
22. Koestler D.J., Robin M.B. Method of producing uniform polymer beads. US 3922255. Nov. 25, 1975.

23. Timm E.E. Process and apparatus for preparing uniform size spheroidal polymer beads and suspension polymerization product thus obtained. EP0051210. May 12, 1982.

24. Timm E.E., Leng D.E. Process for preparing uniformly sized polymer particles by suspension polymerization of vibratorily excited monomers in a gaseous or liquid stream. US 4623706. Nov. 18, 1986.

25. Ikkai F., Shibayama M. Microstructure and swelling behavior of ion-exchange resin. Journal of Polymer Science Part B: Polymer Physics. V. 34 № 9 (1996) pp. 1637-1645.

26. Komers R., Tomanová D. Untersuchung der Textureigenschaften einiger macromoleculärer Harze. Chemicky prumysl. V. 21 № 46 (1971) pp. 559-563.

27. Sugo T., Ishigaki I., Fujiwara K., Sekiguchi H., Kawazu H., Saito T. Electrically regenerable demineralizing apparatus. US 5308467. May 3, 1994.

28. Kankkunen T. Controlled transdermal drug delivery by iontophoersis and ion-exchange fiber. Dissertation, University of Helsinki, 2002.

29. Lilja J., Aumo J., Salmi T., Murzin D.Yu.; Mäki-Arvela P., Sundell M., Ekman K., Peltonen R., Vainio H. Kinetics of esterification of propanoic acid with methanol over a fibrous polymer-supported sulphonic acid catalyst. Applied catalysis A, General. V. 228 № 1-2 (2002) pp. 253-267.

30. Lilja J., Murzina E., Grénman H., Vainio H., Salmi T., Murzin D.Yu. The selective sorption of solvents on sulphonic acid polymer catalyst in binary mixtures. Reactive and Functional Polymers. V. 64 № 2 (August 2005) pp. 111-118.

31. Helfferich F. Ionenaustauscher. Verlag Chemie GmbH, Weinheim, 1959.

32. Helfferich F. Ion Exchange. 621 p. McGraw-Hill, New York. 1962.

33. Gnusin N.P., Grebenyuk V.D., Pevnitskaya M.V. Electrochemistry of granulated ion-exchangers. 200 p. Novosibirsk: Nauka. 1972. (in Russian)

34. Nikolaev N.I. Diffusion in membranes. 384 p. Moscow. Chemistry. 1980.

35. Spitsyn M.A., Andreev V.N., Kazarinov V.E., Kokoulina D.V., Krishtalik L.I. Relation of diffusive and migrative mobilities of sodium ions in perfluorinated sulfocationic membranes. Elektrokhimiya. V. 20 № 8 (1984) pp. 1063 - 1068. (in Russian)

36. Spitsyn M.A., Krishtalik L.I. Relation of diffusive and migrative mobilities of sodium ions in polystyrene-sulfocationic membranes. Elektrokhimiya. V. 21 № 8 (1985) pp. 1133 - 1135. (in Russian)

37. Spitsyn M.A., Krishtalik L.I. About influence of transporting ion on ratio of its diffusive and migrative mobilities in cation-exchange membranes. Elektrokhimiya. V. 23 № 8 (1988) pp. 380-383. (in Russian)

38. Ooi K., Katoh H., Sonoda A., Hirotsu T. Screening of Adsorbents for Boron in Brine. Journal of Ion Exchange. V. 7 № 3 (1996) p. 166.

39. Tanabe M., Kaneko S. High-purity water producing apparatus utilizing boron-selective ion exchange resin. US 5833846. Nov. 10, 1998.

40. Spiegler K.S., Yoest R.L., Wyllie M.R.J. Electrical potentials across porous plugs and membranes. Ion-exchange resin-solution systems. Discussions of the Faraday Society. V. 21 (1956) p. 174.

41. Mahmoud A., Muhr L., Grévillot G., Valentin G., Lapicque F. Ohmic drops in the ion-exchange bed of cationic electrodeionisation cells. Journal of applied electrochemistry. V. 36 (2006) pp. 277-285.

42. Pashkov A.B., Zhukov M.A., Segal T.N., Tikhomirova N.S., Serenkov V.I., Solodkin L.S., Zagorskaya Z.G., Saldadze K.M., Klimova Z.V., Balova N.G. Method of producing homogeneous ion-exchange membranes. GB 1106946. March 20, 1968.

43. Bourat G. Ion-exchange membranes. US 3563921. Feb. 16, 1971.

44. Mizutani Y., Yamane R., Ihara H., Motomura H. Studies of Ion Exchange Membranes. XVI. The Preparation of Ion Exchange Membranes by the "Paste Method". Bulletin of the Chemical Society of Japan. V. 36 № 4 (1963) pp. 361-366.

45. Mizutani Y., Yamane R., Motomura H. Studies of Ion Exchange Membranes. XXII. Semicontinuous Preparation of Ion Exchange Membranes by the "Paste Method". Bulletin of the Chemical Society of Japan. V. 38 № 5 (1965) pp. 689-694.

46. Aminabhavi T., Kulkarni P.V., Kariduraganavar M.Y. Ion-exchange membranes, methods and processes for production thereof and uses in specific applications. US 6814865. Nov. 9, 2004.

47. Sata T. Recent trends in ion exchange membrane research. Pure and Applied chemistry. V. 58 № 12 (1986) pp. 1613-1626.

48. Choi Y.-J., Kang M.-S., Moon S.-H. A new preparation method for cation-exchange membrane using monomer sorption into reinforcing materials. Desalination. V. 146 (2002) pp. 287–291.

49. Wyllie M.R.J., Patnode H.W. The development of membranes prepared from artificial cation-exchange materials with particular reference to the determination of sodium-ion activities. The Journal of Physical Chemistry. V. 54 (1950) pp. 204-227.

50. Grigorov O.N., Kozmina Z.P., Markovitsch A.V., Fridrichsberg D.A. Elecrtochemical properties of capillary systems. Moskau: Nauka. 1956.

51. Saldadze K.M., Pashkov A.B., Titov V.S. Ion-exchange high-molecular substances. Moskau. Nauka. 1967.

52. Gierke T.D., Munn C.E., Walmsley P.N. The morphology in Nafion perfluorinated membrane products, as determined by wide- and small-angle X-ray studies. Journal of Polymer Science, Polym. Phys. Ed. V. 19 (1981) pp. 1687-1704.

53. Heitner-Wirguin C. Recent advances in perfluorinated ionomer membranes - structure, properties and applications. Journal of Membrane Science. V. 120 № 1 (1996) pp. 1-33.

54. Stauffer D., Aharony A. Introduction to Percolation Theory, 2nd ed., Taylor and Francis. 1994. ISBN: 978-0748402533

55. Grimmett G. Percolation, 2nd ed. Springer. 1999. ISBN: 978-3-540-64902-1.

56. Strathmann H. Ion-exchange membrane separation processes. 360 p. Elsevier: Membrane science and technology series, 9. 2004. ISBN: 978-0-444-50236-0.

57. Strathmann H. Membranes and membrane separation processes. In: Ullmann's Encyclopedia of industrial chemistry. 2005.

58. Dyck A. Entwicklung von Membranmaterialien auf Basis aromatischer sulfonierter Polymere und deren Charakterisierung für die Anwendung in Direkt-Methanol-Brennstoffzellen. 152 p. Shaker Verlag. 2003. ISBN: 978-3-8322-1150-9.

59. Karpenko L.V., Demina O.A., Dvorkina G.A., Parshikov S.B., Larchet C. Comparative study of methods used for the determination of electroconductivity of ion-exchange membranes of Electroconductivity of Ion-Exchange Membranes. Russian Journal of Electrochemistry. V. 37 № 3 (2001) pp. 287–293. Translated from Elektrokhimiya, V. 37 № 3 (2001) pp. 328–335.

60. Idelchik I.E. Handbook of hydraulic resistance. CRC Press, 3rd Edition. 1994 ISSN 1-56700-074-6.

61. Gaiser G. Strömungs- und Transportvorgänge in gewellten Strukturen, Dissertation, Universität Stuttgart, 1990.

62. Li F. Novel spacers for membrane filtration processes. Dissertation, Universität Twente, 2003.

63. Zabolotsky V.I., Nikonenko V.V. Ion transport in membranes. Moscow: Nauka. 1996.

64. Tanaka Y. Limiting current density of an ion-exchange membrane and of an electrodialyzer. Journal of membrane science. V. 266 (2005) pp. 6-17.

65. Nikonenko V., Lebedev K., Manzanares J.A., Pourcelly G. Modelling the transport of carbonic anions through anion-exchange membranes. Electrochimica Acta. V. 48 (2003) pp. 3639-3650.

66. Rubinstein I., Zaltzman B. Electro-convective versus electroosmotic instability in concentration polarization. Advances in Colloid and Interface Science. V. 134-135 (2007) pp. 190-200.

67. Rubinstein I. Theory of concentration polarization effects in electrodialysis on counter-ion selectivity of ion-exchange membranes with differing counter-ion distribution coefficients. Journal of the Chemical Society, Faraday Transactions. V. 86 № 10 (1990) pp. 1857-1861.

68. Rubinstein I., Maletzki F. Electroconvection at an electrically inhomogeneous permselective membrane surface. Journal of the Chemical Society, Faraday Transactions. V. 87 № 13 (1991) pp. 2079-2087.

69. Mishchuk N.A. Electro-osmosis of the second kind near the heterogeneous ion-exchange membrane. Colloids and Surfaces A: Physicochemical and Engineering Aspects. V. 140 № 1-3 (1998) pp. 75-89.

70. Volodina E., Pismenskaya N., Nikonenko V., Larchet C., Pourcelly G. Ion transfer across ion-exchange membranes with homogeneous and heterogeneous surfaces. Journal of Colloid and Interface Science. V. 285 № 1 (2005) pp. 247-258.

71. Krol J.J., Wessling M., Strathmann H. Concentration polarization with monopolar ion exchange membranes: current-voltage curves and water dissociation. Journal of membrane science. V. 162 (1999) pp. 145-154 and pp. 155-164.

72. Zabolotsky V.I., Nikonenko V.V., Pismenskaya N.D., Laktionov E.V., Urtenov M.Kh., Strathmann H., Wessling M., Koops G.H. Coupled transport phenomena in overlimiting current electrodialysis. Separation and Purification Technology. V. 14 (1998) pp. 255-267.

73. Neubrand W. Modellierung und Simulation von Electromembranverfahren, Ph.D.Thesis, Universität Stuttgart, Institut für Chemische Verfahrenstechnik, Germany, 1999.

74. Davis T.A., Grebenyuk V.D., Grebenyuk O.V. "Electromembrane processes" in: Membrane technology in the chemical industry. Edited by S.P. Nunes and K.-V. Peinemann. Wiley-VCH. Weinheim. 2001.

75. Pevnitskaya M.V., Kozina A.A, Evseev N.G. Electroosmotic permeability of ion-exchange membranes. Izvestiya. SO AN USSR. Ser. Khimicheskaya. № 4 (1974) pp. 137 – 141.

76. Shel'deshov N.V., Chaika V.V., Zabolotskii V.I. Structural and mathematical models for pressure-dependent electrodiffusion of an electrolyte through heterogeneous ion-exchange membranes: pressure-dependent electrodiffusion of NaOH through the MA-41 anion-exchange membrane. Russian Journal of Electrochemistry. V. 44 № 9 (2008) pp. 1036-1047.

77. Strathmann H., Rapp H.-J., Bauer B., Bell C.-M. Theoretical and practical aspects for preparing bipolar membranes. Desalination. V. 90 (1993) pp. 303-323.

78. Zabolotskii V.I., Sheldeshov N.V., Gnusin N.P. Dissociation of water molecules in systems with ion-exchange membranes. Russian chemical reviews. V. 57 (1988) pp. 1403-1414.

79. Simons R. Electric field effects on proton transfer between ionizable groups and water in ion exchange membranes. Electrochimica Acta. V. 29 (1984) pp. 151–158.

80. Simons R. Strong electric field effects on proton transfer between membrane-bound amines and water. Nature. V. 280 (1979) pp. 824–826.

81. Rubinstein S.M., Manukyan G., Staicu A.D., Rubinstein I., Zaltzman B., Lammertink R.G.H., Mugele F., Wessling M. Direct observation of a nonequilibrium electro-osmotic instability. Physical Review Letters. V. 101 № 23 (2008) p. 236101.

82. Juda W. Construction of cells for Electrodialysis. US 2741595. Apr. 10, 1956.

83. Karn W. Spiral wound electrodialysis cell. US 4225413. Sept. 30, 1980.

84. Giuffrida A.J., Parsi E.J. Method for preventing scale buildup during electrodialysis operation. US 3341441. Sept. 12, 1967.

85. Shaposhnik V.A. Kinetic of electrodialysis. Voronezh: VGU. 1989.

86. „Water Desalination" by Jorge A. Arroyo in "Texas Water Development Board. Report 360 - Aquifers of the Edwards Plateau" edited by Robert E. Mace, Edward S. Angle, and William F. Mullican, February 2004.

87. Krol J.J., Jansink M., Wessling M., Strathmann H. Behaviour of bipolar membranes at high current density: Water diffusion limitation. Separation and Purification Technology. V. 14 (1998) pp. 41-52.

88. Aritomi T., Boomgaard T., Strathmann H. Current-voltage curve of a bipolar membrane at high current density. Desalination. V. 104 (1996) pp. 13-18.

89. Handbook on bipolar membrane technology. Editor: Kemperman A.J.B. Twente University Press. 2000. ISBN 9036515203.

90. Strathmann H. Bipolar Membranes and Membrane Processes. In "Encyclopedia of separation science" V. 2 Membrane separations. Elsevier. 2000.

91. Bazinet L. Electrodialytic phenomena and their applications in dairy industry: a review. Critical reviews in food science and nutrition. V. 45 (2005) pp. 307-326.

92. Neville M.D., Jones C.P., Turner A.D. The EIX process for radioactive waste treatment. Progress in nuclear energy. V. 32 № 3-4 (1998) pp. 397-401.

93. Nyberg E.D. Electrochemically assisted ion exchange. WO 9832525. 1998-07-30.

94. Nyberg E.D., Vogdes C.E., Holmes J.C., Janah A.K. Electrochemical ion exchange treatment of fluids. WO 2007044609. 2007-04-19.

95. Spiegler K.S., Coryell C.D. Electromigration in an cation-exchange resin. II Detailed analysis of two-component systems. The Journal of Physical Chemistry. V. 56, p. 106, 1952.

96. Strathmann H., Kock K. Effluent free electrolytic regeneration of ion-exchange resins. Polymer separation Media, Ed. Cooper A.R. New York: Plenum Press. 1982.

97. Singh R.K., Bajpai D.D., Venugopalan A.K., Divakar D.S., Singh R.R., Patil C.B. Novel method for concentration of low level radioactive waste: Recent advances in membrane-based separation science and technology. Indian Journal of chemical technology. V. 3 (1996) №. 3, pp. 149-151.

98. Wood C.J., Bradbury D. Process for removing radioactive burden from spent nuclear reactor decontamination solutions using electrochemical ion exchange. US 5078842. Jan. 7, 1992.

99. Liu Y., Avdalovic N., Small H. Electrolytic suppressor and separate eluent generator combination. WO 2002004940. Jan. 17, 2002.

100. Srinivasan K., Avdalovic N. Aqueous stream purifier and method of use. WO 03059822. July 24, 2003.

101. Xu T. Ion exchange membranes: State of their development and perspective. Journal of membrane science. V. 263 (2005) pp. 1-29.

102. Frenzel I., Holdik H., Stamatialis D.F., Pourcelly G., Wessling M. Chromic acid recovery by electro-electrodialysis I. Evaluation of anion-exchange membrane. Journal of Membrane Science. V. 261 № 1-2 (2005) pp. 49-57.

103. Frenzel I., Holdik H., Stamatialis D.F., Pourcelly G., Wessling M. Chromic acid recovery by electro-electrodialysis: II. Pilot scale process, development, and optimization. Separation and Purification Technology. V. 47 № 1-2 (2005) pp. 27-35.

104. Luo G.S., Pan S., Liu J.G. Use of the electrodialysis process to concentrate a formic acid solution. Desalination. V. 150 №3 (2002) pp. 227-234.

105. Kaoru Onuki, Gab-Jin Hwang, Arifal, Saburo Shimizu. Electro-electrodialysis of hydriodic acid in the presence of iodine at elevated temperature. Journal of Membrane Science. V. 192 № 1–2 (2001) pp. 193–199.

106. Akgemci E.G., Ersöz M., Atalay T. Applications of electro-electrodialysis for recovery of acids from pickle waste in leather inductry. First international symposium on process intensification and miniaturization. University of Newcastle, August 18 - 21, 2003. p. 67.

107. Huang C., Xu T., Zhang Y., Xue Y., Chen G. Application of electrodialysis to the production of organic acids: state-of-the-art and recent developments. Journal of membrane science. V. 288 (2007) pp. 1-12.

108. Leea Jae-Bong, Parka Kwang-Kyu, Euma Hee-Moon, Leeb Chi-Woo Desalination of a thermal power plant wastewater by membrane capacitive deionization. Desalination. V. 196 (2006) pp. 125–134.

109. Lee J.-B., Park K.-K., Eum H.-M., Lee C.-W. Desalination of a thermal power plant wastewater by membrane capacitive deionization. Desalination. V. 196 (2006) pp. 125-134.

110. Stevens T.S., Jewett G.L., Bredeweg R.A. Liquid chromatographic method and apparatus with a packed tube membrane device for post-column derivatization/suppression reactions. US 4751004. June 14, 1988.

111. Water desalting planning guide for water utilities / Water Desalting Committee, American Water Works Association. 2004. ISBN 0-471-47285-9.

112. Strathmann H, Chmiel H. Die Elektrodialyse – ein Membranverfahren mit vielen Anwendungsmogligkeiten. Chem.-Ing.Tech. V. 56 № 3 (1984) pp. 214-220.

113. Davis T.A. Production of purified water and high value chemicals from salt water. WO 2004013048. Febr. 12, 2004.

114. Sferrazza A., Schmidt E., Williams M.E. Integrated electro-pressure membrane deionization system. WO2006074259. July 13, 2006.

115. Nagel R. Elektrodeionisation (EDI) bei der Zusatzwasseraufbereitung in Kraftwerksanwendungen. VGB PowerTech – International journal for electricity and heat generation. V. 85 № 5 (2005) pp. 112-117.

116. Strathmann H., Grabowski A., Eigenberger G. Electromembrane processes, efficient and versatile tools in a sustainable industrial development. Desalination. V. 199 (2006) pp. 1-3.

117. Ibanez R., Stamatialis D.F., Wessling M. Role of membrane surface in concentration polarization at cation exchange membranes. Journal of Membrane Science. V. 239 (2004) pp. 119–128.

118. Peters A.M., Lammertink R.G.H., Wessling M. Comparing flat and micro-patterned surfaces: gas permeation and tensile stress measurements. Journal of Membrane Science. V. 320 № 1-2 (2008) pp. 173-178.

119. Scott K., Mahmood A.J., Jachuck R.J., Hu B. Intensified membrane filtration with corrugated membranes. Journal of Membrane Science. V. 173 № 1 (2000) pp. 1-16.

120. Scott K., Jachuck R.J., Hall D. Crossflow microfiltration of water-in-oil emulsions using corrugated membranes. Separation and purification technology. V. 22-23 (2001) pp. 431-441.

121. Scott K., Lobato J. Mass transport in cross-corrugated membranes and the influence of TiO₂ for separation processes. Industrial & Engineering Chemistry Research. V. 42 (2003) pp. 5697-5701.

122. Scott K., Lobato J. Mass transfer characteristics of cross-corrugated membranes. Desalination. V. 146 (2002) pp. 255-258.

123. Hall D.W., Scott K., Jachuck R.J. Determination of mass transfer coefficient of a cross-corrugated membrane reactor by the limiting-current technique. International journal of heat and mass transfer. V. 44 (2001) pp. 2201-2207.

124. Tzanetakis N., Taama W.M., Scott K., Jachuck R.J.J. The effect of corrugated membranes on salt splitting. Journal of Applied Electrochemistry. V. 33 № 5 (2003) pp. 411-417.

125. Tzanetakis N., Taama W.M., Scott K., Jachuck R.J.J., Slade R.S., Varcoe J. Comparative performance of ion exchange membranes for electrodialysis of nickel and cobalt. Separation and Purification Technology. V. 30 № 2 (2003) pp. 113-127.

126. Van der Waal M.J., and Racz I.G. Mass transfer in corrugated-plate membrane modules. I. Hyperfiltration experiments. Journal of Membrane Science. V. 40 № 2 (1989) pp. 243-260.

127. Gschwind P., Regele A., Kottke V. Sinusoidal wavy channels with Taylor-Goertler vortices. Experimental Thermal and Fluid Science. V. 11 № 3 (1995) pp. 270-275.

128. Zimmerer C., Gschwind P., Gaiser G., Kottke V. Comparison of heat and mass transfer in different heat exchanger geometries with corrugated walls. Experimental thermal and fluid science. V. 26 № 2-4 (2002) pp. 269-273.

129. Zhang L.-Z. Convective mass transport in cross-corrugated membrane exchangers. Journal of membrane science. V. 260 № 1-2 (2005) pp. 75-83.

130. Zabolotskii V.I., Loza S.A., Sharafan M.V. Physicochemical properties of profiled heterogeneous ion-exchange membranes. Russian journal of electrochemistry. V. 41 № 10 (2005) pp. 1053–1060. Translated from Elektrokhimiya, V. 41 (2005) №. 10 pp. 1185–1192.

131. Polyanskii N.G., Gorbunov G.V., Polyanskaya N.L. Metody issledovaniya ionitov. 208 p. Moscow: Khimiya. 1976. (In Russian)

132. Oren Y., Freger V., Linder C. Highly conductive ordered heterogeneous ion-exchange membranes. Journal of Membrane Science. V. 239 (2004) pp. 17-26.

133. Liu D., Yates M.Z. Tailoring the structure of S-PEEK/PDMS proton conductive membranes through applied electric fields. Journal of Membrane Science. V. 322 (2008) pp. 256–264.

134. Walters W.R, Weiser D.W., Marek L.J Concentration of radioactive aqueous wastes Electromigration through ion-exchange membranes. Industrial and engineering chemistry. V. 47 (1955) pp. 61-67.

135. Kollsman P. Method and apparatus for treating ionic fluids by dialysis. US 2815320. Dec. 3, 1957.

136. Process and apparatus for the electrolytic deionisation of salt-containing liquids. GB776469. Jun. 5, 1957. Nederlandse Organisatie Voor Toegepast - Natuurzetenschappelijiz Onderzoer Ten Behoeve Vannzjverheid.

137. Tye F.L. Improvements relating to electrodialysis processes. GB 815154, June 17, 1959.

138. Kressman T.R.E. Process for the removal of dissolved solids from liquids. US 2923674. Febr. 2, 1960.

139. Glueckauf E. Electro-deionization through a packed bed. British Chemical Engineering. V. 4 (December 1959) pp. 646-651.

140. Gittens G.J., Glueckauf E. The application of electrodialysis to demineralization. AIChE IChemE Symposium Series № 9, London (1965) pp. 79-83.

141. Gittens G.J., Watts R.E. Some experimental studies of electrodeionization through resin packed beds. United Kingdom Atomic Energy Authority Research Report, 1964.

142. http://www.cediuniversity.com/

143. Matějka Z. Continuous and simultaneous removal of anions and dissolved oxygen from water by an ion exchange membrane cell. Bulletin of the chemical society of Japan. V. 41. № 4 (1968) pp. 1024-1025.

144. McRay W.A., Rigopulos P.N. Removal of gases by electrode-ionization. US 3330750. July 11, 1967.

145. Matějka Z. Continuous production of high purity water by electro-deionisation. Journal of applied chemistry and biotechnology. V. 21 (1971) pp. 117-120.

146. Shaposhnik V.A., Reshetnikova A.K., Zolotareva R.I., Drobysheva I.V., Isaev N.I., Demineralization of water by electrodialysis with ion-exchanger packing between the membranes. Zhurnal Prikladnoi Khimii. V. 46 №12 (1973) pp. 2659-2663. (in Russian)

147. Shaposhnik V.A., Reshetnikova A.K., Kobeleva N.S. Choice of the desalination path length in the production of ultrapure water by electrodeionization. Zhurnal prikladnoi khimii. V. 53 №2 (1980) pp. 443-445.

148. Pevnitskaya M.V., Starikovskiy L.G., Usov V.J., Borodikhina L.I. Investigation of electrodeionization apparatus by deionisation of water and process optimisation. Zhurnal prikladnoi khimii. V. 54 № 9 (1981) pp. 2077-2081.

149. Guiffrida A.J., Jha A.D., Ganzi G.C. Electrodeionization apparatus. US 4632745. Dec. 30, 1986.

150. Ganzi, G.C., Egozy, Y., Giuffrida, A.J., Jha, A.D. High purity water by electrodeionization Performance of the Ionpure™ continuous deionization system. Ultrapure Water V. 4 № 3 (April 1987) pp. 43-50.

151. Liang L.-S. Evolution in design of CEDI systems. Ultrapure water. V. 20 №8 (October 2003) pp. 13-19.

152. Bennett A. Advances in high purity water filtration technologies. Filtration+Separation. V. 41 (2004) pp. 28-30.

153. Fedorenko V.I. Ultrapure water production by continuous electrodeionization method: technology and economy. Pharmaceutical Chemistry Journal. V. 38 (2004) pp. 35-40.

154. Bergmann H., Iourtchouk T., Rittel A., Zuleeg H. New environmental orientated electro-membrane technologies for industrial process and waste water // Achema 2006. 28th International Exhibition-Congress on Chemical Engineering, environmental protection and biotechnology. Frankfurt am Mein, 15-19 May 2006.

155. Johann J. Elektrodialytische Regenerierung von Ionenaustauscherharzen, Ph. D. Thesis, Universität Stuttgart, Institut für Chemische Verfahrenstechnik, Stuttgart, 1992.

156. Yeong K.-H., Moon S.-H. A study on removal of cobalt from a primary coolant by continuous electrodeionization with various conducting spacers. Separation science and technology. V. 38 (2003) pp. 2347-2371.

157. Yeong K.-H., Song J.-H., Moon S.-H. A study on stack configuration of continuous electrodeionization for removal of heavy metal ions from the primary coolant of a nuclear power plant. Water research. V. 38 (2004) pp. 1911-1919.

158. YuPo J. Lin, M. Henry, J. Hestekin, S. W. Snyder, E. J. St. Martin Single stage separation and esterification of cation salt carboxylates using electrodeionization. US 2005/0056547 A1 Mar. 17, 2005.

159. Arora M.B., Hestekin J.A., Lin YuPo J., Martin E.J.St., Snyder S.W. Immobilized biocatalytic enzymes in electrodeionization. US 2004/00115783 A1. Jun.17, 2004.

160. YuPo Lin, Michael Henry, Seth Snyder Applications of resin-wafer electrodeionization and the separative bioreactor in biorefineries. NAMS 2006, May 12-17, 2006. Chicago, USA

161. Kedem O., Maoz Y. Ion-conducting spacer for improved Electrodialysis. Desalination. V. 19 (1976) pp. 465-470.

162. Dejean E., Laktionov E., Sandeaux J., Sandeaux R., Pourcelly G., Gavach C. Electrodeionization with ion-exchange textile for the production of high resistivity water: Influence of the nature of the textile. Desalination. V. 114 № 2 (1997) pp. 165-173.

163. Laktionov E., Dejean E., Sandeaux J., Sandeaux R., Gavach C., Pourcelly G. Production of high resistivity water by electrodialysis. Influence of Ion-Exchange Textiles as Conducting Spacers. Separation Science and Technology. V. 34 № 1 (1999) pp. 69-84.

164. Mir L. Electrodeionization apparatus with fixed ion exchange materials. US 6241866. June 5, 2001.

165. Myriam Ben Chehida Elleuch, Mohamed Ben Amor, Gerald Pourcelly Phosphoric acid purification by a membrane process: Electrodeionization on ion-exchange textiles. Separation and Purification Technology. V. 51 № 3 (October 2006) pp. 285-290.

166. Sue Hyung Choi, Seung-Hyeon Moon, Man Bock Gu Biodegradation of chlorophenols using the cell-free culture broth of Phanerochaete chrysosporium immobilized in polyurethane foam. J. Chem. Technol. Biotechnol. V. 77 (2002) pp. 999-1004.

167. Kyeong-Ho Yeon, Jin-Woo Lee, Jae-Suk Lee and Seung-Hyeon Moon Preparation and characterization of cation exchange media based on flexible polyurethane foams. J. Appl. Poly. Sci. V. 86 (2002) pp. 1773-1781.

168. Inoue H., Yamanaka K., Yoshida A., Aoki T., Teraguchi M., Kaneko T. Synthesis and cation exchange properties of a new porous cation exchange resin having an open-celled monolith structure. Polymer. V. 45 (2004) pp. 4-7.

169. Datta R., Lin YuPo, Burke D., Tsai Shih-Perng Electrodeionization substrate, and device for electrodeionization treatment. US 6495014. Dec. 17, 2002.

170. Ganzi G., Wilkins F., Giuffrida A., Griffin C. Electrodeionization apparatus. WO 9211089. July 9, 1992.

171. Rapp H.-J. Die Elektrodialyse mit bipolaren Membranen – Theorie und Anwendung. Dissertation, Institut für Chemische Verfahrenstechnik, Univetsitaet Stuttgart, 1995.

172. Darbouret D., Kano I. Ultrapure water blank for boron trace analysis. Journal of Analytical Atomic Spectrometry. V. 15 (2000) pp. 1395-1399.

173. Ruimei Wen, Shouquan Deng and Yafeng Zhang The removal of silicon and boron from ultra-pure water by electrodeionization. Desalination. V. 181 № 1-3 (2005) pp. 153-159.

174. Pevnitskaya M.V., Starikovskii L.G., Usov V.Yu., Borodichina L.I. Operation of mixed-bed electrodialysis for exhaustive deionization of water and routes to optimisation of the process. Journal of applied chemistry of the USSR (English translation) V. 54 № 9 part 2 (1981) pp. 1818-1821.

175. Tessier D.F, Toupin J.D.R., Towe I.G. Electrodeionization apparatus having geometric arrangement of ion exchange material. WO 9725147. July 17, 1997.

176. Kunz G.K. "Process and apparatus for treatment of fluids, particularly desalinization of aqueous solutions. US 4636296. 1987.

177. Ganzi G.C., Jha A., DiMascio F. and Wood J. The Theory and Practice of Continuous Electrodeionization. Ultrapure Water Journal. V. 14 № 6 (July/August 1997) pp. 64-69.

178. Kunz G. Verfahren und Vorrichtung zum behandeln von Flüssigkeiten, insbesondere entsalzen wässriger Lösungen. DE 3329813. February 28, 1985.

179. Salem E., Tessier D.F. Advances in Electrodeionization. International Water Conference, Pittsburgh, PA. October 1998.

180. Jha A.D., Gifford J.D. Ultrapure CEDI for microelectronics applications: a cost effective alternative to mixed-bed polishers. Proceedings for Ultrapure water Asia 2004, Singapore, March 10-11, 2004.

181. Franzreb M. Device for magnetically controlled electrodeionization. WO 03028863, Apr. 10, 2003.

182. Lacher C., Franzreb M., Höll W. H. Improving the efficiency of electrodeionization by means of magnetic ion exchange resins. PowerPlant Chemistry. V. 6 № 8 (2004).

183. Lacher C., Franzreb M., Höll W.H. Electrodeionization using magnetic ion exchange resins. Euromembrane 2004, 28 Sept.-1 Oct. 2004, Hamburg. Poster S8-P-27 and "Book of abstracts" p. 226.

184. Parsi E. Removal of dissolved saltsand silica from liquids. US 3149061. Sept. 15, 1964.

185. E. Korngold, E. Selegny Method of separation of ions from a solution. US 3686089. August 22, 1972.

216

186. Kunz G. Verfahren und Vorrichtung zum Behandeln von Flüssigkeiten, insbesondere Entsalzen wässriger Lösungen. DE 3217990. Nov. 17, 1983.
187. Neumeister H., Fürst L., Flucht R. Electrolytic cell arrangement for the deionization of aqueous solutions. US 5954935. Sept. 21, 1999.
188. Parsi E.J. Apparatus for the removal of dissolved solids from liquids using bipolar membranes, US 4871431. Oct. 3, 1989.
189. Parsi E. Apparatus and process for the removal of acidic and basic gases from fluid mixtures using bipolar membranes. US 4969983. Nov. 13, 1990.
190. Neumeister H., Fürst L., Flucht R. Einfach und Mehrfachelektrolysezellen sowie Anordnungen davon zur Entionisierung von wässrigen Medien. DE 4418812 A1, Dec. 7, 1995.
191. Neumeister H., Flucht R., Fürst L., Nguyen V.D., Verbeek H.M. Theory and experiments involving an electrodeionization process for high-purity water production. Ultrapure water. V. 17 (April 2000) pp. 22-30.
192. Thate S., Specogna N., Eigenberger G. A comparison of different EDI concepts used for the production of high-purity water. Ultrapure water. V. 16 № 8 (October 1999) pp. 42-56.
193. Thate S. Untersuchung der elektrochemischen Deionisation zur Reinstwasserherstellung, Ph.D.Thesis, Universität Stuttgart, Institut für Chemische Verfahrenstechnik, Stuttgart, Germany (February 2002).
194. Parsi E.J. Removal of weakly basic substances from solution by electrodeionization. US 3291713. Dec. 13, 1966.
195. Ganzi G.C., Jha A.D., DiMascio F., Wood J.H. Electrodeionization – Theory and practice of continuous electrodeionization. Ultrapure water. V. 14 № 6 (July/August 1997) pp. 64-69.
196. Gifford J.D. Atnoor D. An innovative approach to continuous electrodeionization module and system design for power applications. Proceedings of Industrial Water Conference, Pittsburg, Pa. (October 2000).
197. Parsi E.J., Sims K.J., Elyanow I.D., Prato T.A. Introducing and removing ion-exchange and other particulates from an electrodeionization stack. US 5066375. Nov. 19, 1991.
198. Towe I.G., Tessier D.F., Huehnergarrd M.P. Modular apparatus for demineralization of liquids. US 6193869 B1, February 27, 2001.
199. Proulx A.G. Membrane module assembly. US 5681438. Oct. 28, 1997.
200. Rychen P., Alonso S., Alt H., Gensbittel D. Apparatus for continuous electrochemical desalination of aqueous solutions. US 5376253. December 27, 1994.
201. Rychen P., Leet J. Electrodeionization The use of EDI in treating semiconductor grade water. Ultrapure water. (Febr. 2000) pp. 36-41.
202. Tate J. Add polish to high purity water with EDI. Water technology magazin. (August 2000).
203. The industry standard in chemical-free water treatment. E-Cell brochure 2001.
204. Liang L., Wang L. Continuous electrodeionization processes for production of ultrapure water. Semiconductor Pure Water and Chemical Conference (2001).
205. Giuffrida A.J., Ganzi G.C., Oren Y. Electrodeionization apparatus and module. US 4956071. Sept. 11, 1990.
206. Denoncourt J.P., Moulin J. Electrodeionization process for purifying a liquid. US 5593563. Jan. 14, 1997.
207. Castillo E., Coleman D.E., Darbouret D., Dimitrakopoulos T., Feuillas E., Vanatta L.E. Qualification of an electrodeionization module via experimental design and ion chromatographic studies. Journal of chromatography A. V. 1039 (2004) pp. 63-70.

208. Thate S., Specogna N., Eigenberger G. Production of high-purity water by continuous electrodeionization with bipolar membranes. Proceedings Euromembrane 2000 Conference, Jerusalem, September 25-27 (2000) pp. 102-103.

209. Thate S., Specogna N., Eigenberger G. Modelling and simulation of high-purity water production using bipolar membranes. Proceedings Conference on Electro-Membrane Processes and Bipolar Membrane Technology, University of Twente, October 25-27 (2000) pp. 92-97.

210. Thate S., Eigenberger G. Untersuchung der elektrochemischen Deionisation zur Reinstwasserherstellung. Vom Wasser. V. 101 (2003) pp. 243-248.

211. Song Jung-Hoon, Yeon Kyeong-Ho, Cho Jaeweon and Moon Seung-Hyeon Effects of the Operating Parameters on the Reverse Osmosis-Electrodeionization Performance in the Production of High Purity Water. Korean Journal of Chemical Engineering. V. 22 №1 (2005) pp. 108-114.

212. Grabowski A., Zhang G., Strathmann H., Eigenberger G. Influence of the anion-exchange membrane permselectivity on the production of high purity water by continuous electrodeionization with bipolar membranes. Journal of membrane science. V. 281 (2006) pp. 297-306.

213. Aristov I.V., Bobreshova O.V., Balavadze E.M. The influence of hydrostatic pressure on the selectivity of electromembrane systems under the limiting conditions of the concentration polarization. Russian Journal of Electrochemistry. V. 32 (1996) pp. 1029-1032.

214. Bobreshova O., Aristov I., Kulintsov P., Balavadze E. Interfacially driven ionic transport in the electromembrane systems under influence of small excess of hydrostatic pressure. Journal of Membrane Science. V. 177 (2000) pp. 201-206.

215. Bobreshova O.V., Kulintsov P.I., Aristov I.V., Balavadze E.M. Selectivity of ion-exchange membranes as a function of hydrostatic pressure. Russian Journal of Electrochemistry. V. 32 (1996) pp. 164-166.

216. Sheldeshov N.V., Chaika V. V., Zabolotskii V.I. Structural and mathematical models for pressure-dependent electrodiffusion of an electrolyte through heterogeneous ion-exchange membranes: pressure-dependent electrodiffusion of NaOH through the MA-41 anion-exchange membrane. Russian Journal of Electrochemistry. V. 44 № 9 (2008) pp. 1036-1046.

217. Zimmerer C.C., Kottke V. Effects of spacer geometry on pressure drop, mass transfer, mixing behavior, and residence time distribution. Desalination. V. 104 № 1-2 (April 1996) pp. 129-134.

218. Ervan Y., Wenten I.G. Study on the influence of applied voltage and feed concentration on the performance of electrodeionization. Songklanakarin J. Sci. Technol. V. 24 (Suppl.) (2002) pp. 955-963.

219. Wang J., Wang S., Jin M. A study of the electrodeionization process - high purity water production with a RO/EDI System. Desalination. V. 132 (2000) pp. 349-352.

220. Mabic S., Kano I., Darbouret D. Ultrapure water for elemental analysis (Poster - P04) Millipore - Technical Library, May 2006.

221. Barber J. Method and apparatus for shifting current distribution in electrodeionization systems. WO 2007/143296. December 13, 2007.

222. International Technology Roadmap for Semiconductors. 2006 Update yield enhancement.

223. http://www.elgalabwater.com/?id=502

224. ASTM D 1193, "Standard Specification for Reagent Water," ASTM International.

225. On-line measurement of dissolved organics in water. Galvanic Applied Sciences. http://www.galvanic.com/UV140.pdf

Curriculum Vitae

Name:	Andrej Grabowski	
Date of Birth:	29. June 1974	
Place of Birth:	Nowopokrowskaja, Russia	
Nationality:	German / Russian	
Marital status:	Married, 2 Children	

School Education	1980 - 1984	Primary School № 7, Nowopokrowskaja (Russia)
	1984 - 1991	School № 10, Nowopokrowskaja (Russia)
Study	1991 - 1996	Faculty of Chemistry, Kuban State University, Krasnodar (Russia)
Scientific Occupation	1996 - 2002	Research Assistant at the Physical Chemistry Department, Kuban State University (Russia)
	2002 - 2006	Research Assistant at Institute for Chemical Processes Engineering, Universität Stuttgart
Industrial Experience	2007 - dato	Senior Scientist, Research and Development, Millipore S.A.S., St. Quentin en Yvelines (France)